NATURE vs NURTURE vs ASTROLOGY

The Only Book To Reveal Scientific Evidence Behind
Astrology, Human Personality, Compatibility & Intelligence

An Illustrated Introduction To

KENEMONICS

K. A. AMON

KENEMONICS

Revealing the Scientific Evidence Behind
Human Personality, Compatibility & Intelligence

An Illustrated Introduction

http://www.kenemonics.net

Twitter @KENEMONICS

Email: info@kenemonics.net

Nature vs Nurture vs Astrology: An Illustrated Introduction to Kenemonics
1st Edition
Revision:000100

Copyright KN Media Publishing LLC 2012-2018 ©
Washington D.C.

ISBN: 978-0-9854149-1-7

This work is partially developed from theories introduced
 in an unreleased title, "Stars & Genetics" 1998, by the same author.

A Deep Gratitude and Appreciation to Aunt Muriel & Professor Hammond

for providing Essential Brilliance & Guidance

CONTENTS

Foreword		6
PART 1 – Astrology & The Zodiac		10
Section 1	Misconceptions	11
Section 2	The Detractors	24
Section 3	Astrology In Everyday Life	28
Section 4	Origin of Astrology	39
PART 2 – Seeing The Connections		45
Section 1	Connecting The Pieces	46
Section 2	The System Of Planets	53
Section 3	The Essential Brain	67
Section 4	Putting it All together	89

PART 3 – FRONTIERS OF MODERN SCIENCE	104
SECTION 1 MODERN SCIENCE	105
SECTION 2 REMAINING MYSTERIES	107
SECTION 3 HUMAN BEHAVIOR	116
SECTION 4 FULL CIRCLE	124
APPENDIX	132
APPENDIX A - ALIGNMENTS & SEISMIC DATA	133
APPENDIX B – OVERLAPPING FIELDS	135
APPENDIX C - FROM TESLA TO NOW	136
BIBLIOGRAPHY	137
INDEX	141

FOREWORD

What makes you unique? What make us different from each other? Many say it's the genes we get from our parents. And how much of a role does our upbringing and early childhood play? For sure nature and nurture both play a part. But not everything is explained by those two. Are we even asking the right questions? Let us start at the most basic level.

The famous scientist Carl Sagan once said, "We are star stuff"* - the product of stars. It is true. Take a look at your body, your hands, arms, skin and bones, you are made of carbon and other elements that come from deep within stars. *In fact, everything on this planet and all other planets come from stars!* Planets are always connected to stars through orbits and cycles of movement, and so are we. As powerful as that statement is, it doesn't go far enough. There is more to this connection besides the obvious light and heat energy the sun gives us. Without those we simply would not exist. Yet since ancient times, as far back as any recorded history, a much more meaningful connection between the celestial bodies and humankind has been observed and studied – today we call that astrology.

The astrology of today is similar to what was practiced by the same ancient society that claimed, "if we lived our lives striving to be the best version of ourselves, when we die we become a star in the sky." Maybe this was just a myth that pharaohs told their children. In a way the ancients hinted that they knew where we came from by telling us it was their highest achievement to return to that source. Never the less, they practiced an astrology that taught a deep connection between humankind and the celestial bodies, one that is much deeper than just mere "star stuff."

This book will explore our mysterious connection to the celestial bodies and show how they affect us. We will start by questioning the claims of astrology, and focusing on the source of human individual personality. Many are familiar with the basic ideas of astrology. Some others dedicate years to practicing and studying it, but no one has been able to explain how it works? This book will.

Hovering over the title of this book is the word, Kenemonics. Kenemonics is a new field studying how humans are affected by celestial bodies. It takes the hidden truths of astrology and combines those with the current understanding of several areas of science, from astronomy to neuroscience, revealing evidence linking the movements of the planets with the subtle structures of the human brain. Even though scientists are not researching for this purpose, it is likely that these discoveries will reveal differences in each person's brain structure that will correlate with the planets positions at the time of birth. So, just like our genes have been proven to be the blueprint for our physical traits, the planets of our solar system provide our basic personality traits, a basic idea passed down by way of Astrology for thousands of years.

The first time astrology was presented to me as a serious subject was in the most unlikely of places, during my time working in a Physics lab as an undergraduate. There was a perfect timing between my mentor, who ran the lab, mentioning the merits of Astrology and a suggestion of my aunt, who was giving me relationship advice. My aunt told me I should get my birth chart done. I asked, "what the heck is a birth chart?". She laughed and explained by making my birth chart. I was amazed. I wasn't completely convinced that astrology was real, but I was impressed by how she described specific details of my personality that only I knew. Still, I wasn't going to blindly accept anything. Nothing was going to escape me analyzing this with the same discipline I used when designing circuits in the electronics lab.

As an engineering undergraduate I was taught how to experiment, observe, measure, and verify if something was actually true. So that's what I did with astrology. The majority of university science silently pushed the idea that "science" was only there for us to learn about the nature of objects, "things" that had no direct and meaningful connection to us as individuals in our daily lives. But astrology is different. One of the fascinating things about astrology is that it is a subject that can be completely about you.

Skeptically, I tested the claims of astrology, and practiced its art by creating hundreds, even thousands of birth charts. It became clear that there was a science about human personality traits "given" to us at birth. I had stumbled upon something real. After years of studying and reading every astrology book I could find, something was still missing – an explanation. No astrologer was able to or attempted to explain how the sun, moon and planets could affect us. That mystery was a spark in my mind, an unsolved puzzle that just might affect all of humankind. That spark ignited a fire and has fueled me ever since. If astrology was true, I would have to study a wide range of sciences in order to uncover its secrets. I also realized that science didn't have to be cold, detached and devoid of personal meaning anymore.

Speaking of cold and detached, many prominent scientists have publicly made false statements attacking the star constellations of the zodiac, attempting to debunk astrology, even spreading misinformation about a "13th sign." For this reason, part one of this book is dedicated to clearing up the popular misconceptions surrounding astrology, exposing the flaws in the arguments of the would-be debunkers of astrology, detailing the actual physical and therefore true basis for astrology and finally covering the time tested principles of astrology.

We live in this wonderful age of high technology. We have smartphones, packed with apps, that we can't seem to put down. Our world is digitally connected, allowing us to get the answer to nearly any question with a quick internet search. Yet modern science cannot explain where our personalities come from or the basis for compatibility. At the same time astrology, which does claim to explain these questions is dismissed outright as nonsense pseudo-science.

Is Astrology total nonsense? Well, let's find out. Fortunately, many can and have seen the very real and distinct individual human personality in each of us. The truth is there. We need to uncover its nature and inner workings. Detractors want a certain type of proof – "hard evidence". They don't see hard evidence right away so they ignore other evidence. The supporters of astrology are satisfied with the evidence they see in people's behavior and don't need any more explanation to be convinced. Another challenge is that personality is much more complex than physical traits like hair color. Even if we can agree that personality traits exists, there are so many factors like opposing traits and life situations that make personality a much more fluid topic. Even so, it can be done with the right approach.

We are going to need three main tools to solve the puzzle of human personality, astrology, the science of the brain and the study of planets. This mix of topics may seem strange and completely unrelated now, but as we dig deeper the connections will become clear. Our search for where personality traits come from is no different than the old search for physical traits that were being passed on from parents to children. Over time, with precise effort we discovered it was genes, given by our parents, that contained the blueprint for these physical traits.

Unfortunately, we currently have an approach to science that ignores a deeper meaning in our daily lives, and denies the connection between all things. In the overall scheme of life, dealing with "things" as separate is not a strength it is a limitation. This narrow approach to the natural world has resulted in a false intellectual arrogance, an arrogance that prevents self-correction and re-evaluation even on a collective level. It prevents the kind of re-evaluation needed to advance us beyond our current level of understanding. True strength lies in seeing the harmony of nature, the connections of everything as parts of the whole.

However, we are still fortunate to live in a time that presents us with the possibility of a great discoveries every day. By the end of this book, you will have a clear view of the true influence that the sun, the moon and the planets have upon our lives. Of course not all questions or mysteries will be answered to the satisfaction of all. However, most of the basic questions will be fully answered and a solid path for future discovery will be established. I wrote this book to inform, but I also hope that it will inspire some young minds to go into fields of science with a fresh approach similar to the excitement and enthusiasm I had as a young student. The true power of information is transformation. Information changes us and allows us to change our world. The information in this book will make a positive and lasting change in the way we see ourselves, both in our world and within the greater cosmos.

PART ONE
ASTROLOGY & THE ZODIAC

By far the single largest misconception of astrology is the true role that the stars play in astrology. While it is true that the *zodiac signs* play a major role in Astrology, the idea that the *"signs"* are the star constellations is a huge misconception. We will discover that astrology, is more closely related to seasons, and is actually based on the interactions between the Earth, the Sun, the Moon and the other planets.

For thousands of years astrology and astronomy were practiced together in ancient societies. Maybe this is the reason causal observers and detractors think of far away stars playing a significant role in astrology. Over time, many details have been lost and fragmented. As a result, clear meanings have been muddled giving rise to many misunderstandings and misconceptions.

Even astrologers do not agree on the role of stars in astrology. There are two schools of thought. One believes the stars are the real zodiac and the other group believes the basis is the equinox. That is the zodiac starts when the sun enters a certain point and the earth's spring equinox begins. The purists push the star idea because everyone can see the stars, they are real. But there are far too many flaws in that argument. Until now no one has given an explanation and a physical source for astrology.

Astrology & The Zodiac 11

    A surprising number of people believe there is some truth to astrology. Out of those people, many have a basic idea about the traits of their sign. And even though most people know their birth sign has something to do with their birthday, they don't know much else. In fact, you could read many books on astrology and the topic of "how it works" may never come up. It's an accepted mystery. To the credit of most authors, many books deal with "the proof," compelling descriptions of personality and compatibility. But is the age-old practice of astrology just a mass delusion? No, it is not, there is something very real behind it. The answer starts with astrology's birth chart.

    The birth chart is basically a chart of planets' positions at the exact time of your birth. Like a huge selfie of the solar system showing the position of the sun, moon and planets. The first step is to separate the lore of astrology from its valid parts. To do this, we have to compare two key things; the star constellations of the zodiac and the birth chart. Below we have the birth chart of Nikola Tesla, it may look strange now but soon you will be familiar with the symbols and layout.

Zodiac Star Constellations fig1

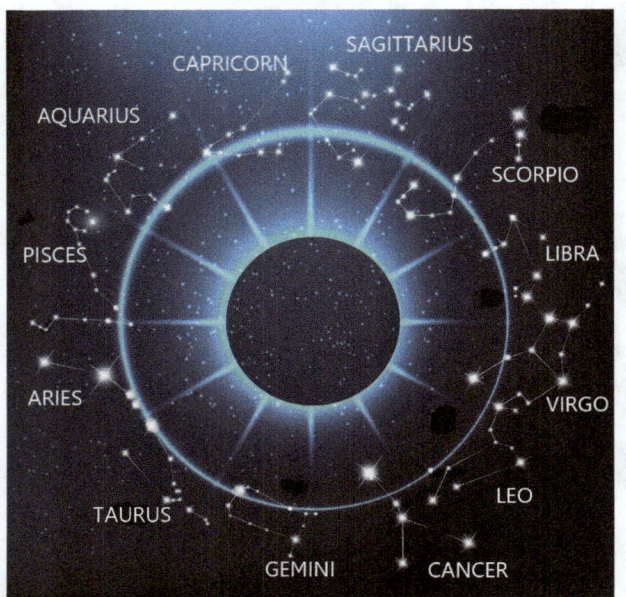

Birth Chart of Nikola Tesla fig2

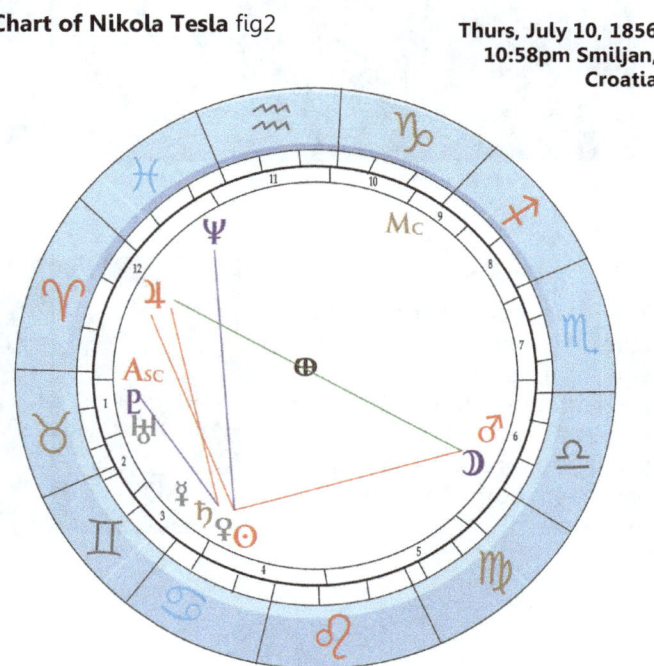

Thurs, July 10, 1856
10:58pm Smiljan,
Croatia

Obviously False

The only similarity the birth chart has with the zodiac constellations is the names of the signs. Astrology does not use the stars at all, and here are the best examples why. The following observations are enough to fuel even casual observers to think astrology is totally invalid, if astrology were based on the stars. In spite of this, astrology has many dedicated supporters including those who venture past these glaring holes and actually create birth charts or those who happen to connect with the accurate description of personalities due to birth signs and the sign compatibilities that can be found in a well written astrology book.

Countless Stars

This would mean even more than just 12 or even 13 constellations. This could never be the basis for Astrology, the choice of stars and arbitrary amount of choices for stars is simply too vast.

Partial Zodiac Ring cross section

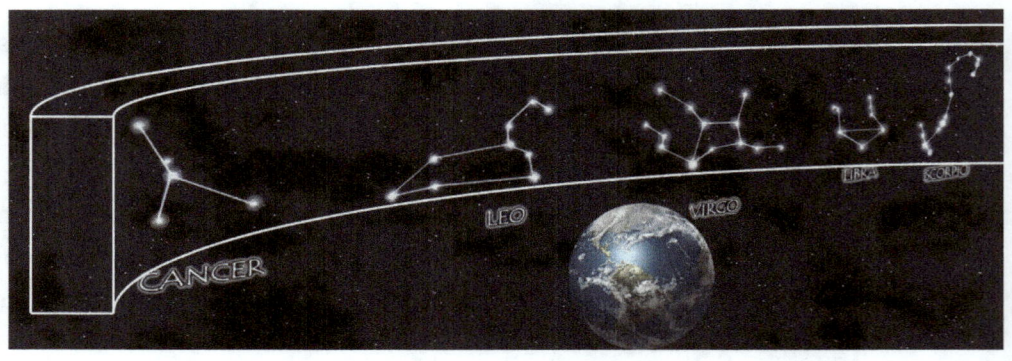

Not 30° per Sign

How could each constellation measure 30 degrees perfectly? Where does one constellation end and the other start given the gaps? Some constellations are large while others are small. This contradicts the precision of the birth chart - 12 signs equally divided at 30° for each sign.

Precession

The star constellations do not currently match up with the birth signs. This is fully known and accounted for during the creation of birth charts. If you go by the stars everyone who is alive now has been born one sign behind their sign.

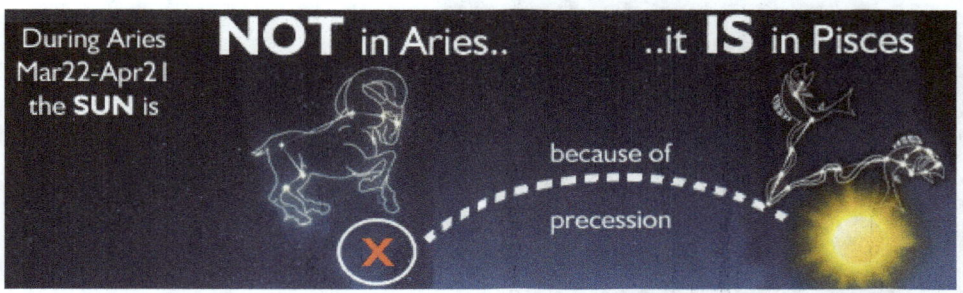

LET US VISUALIZE

If Astrology does not use the stars, but actually uses the spring equinox, then where do the "Zodiac Signs" used in astrology come from? We are going to answer that, but first we need to understand the basics about the Earth and the planets orbiting the Sun. After that we will build an actual birth chart.

The picture to the right shows our basic solar system from a top view. From this view we get a sense of the solar system being a flat disc. That sense is pretty accurate. The planets and the sun are more or less sit on a flat plane, with some planet's orbits going above and below this plane more than others.

Another way to visualize the solar system is to think of balls on a table. This looks simple but there is much more going on, this relatively flat base sets up a very important relationship between the planets.

But the question remains. Where do Astrology's Zodiac signs come from? *The simple answer is… The Tilt of the Earth's Axis.* The bottom picture on this page shows a globe on a table with its tilt. It's well know that this tilt causes the seasons, *but it is virtually unknown that this same tilt gives us the zodiac.*

For the rest of this section, the best way to follow along would be to get a small globe, set it on a table and move it around a stationary object while paying very close attention to the axis of the globe.

WHY DO WE HAVE SEASONS?

We have seasons because of the earth's changing position as it revolves around the sun. That is the simple answer. In summer the north tilts toward the sun making it hot. In winter the north tilts away from the sun making it cold. Throughout the year each position of the Earth around Sun is unique, not because of the stars, but due to the Earth's Axis tilt. The signs of Aries, Cancer, Libra, and Capricorn are called the Cardinal Signs in astrology; the beginnings of the seasons. These yearly positions show the true Zodiac that is used in the birth chart.

All of the other planets including the Moon have this relationship with the Earth's axis,. This relationship is the true basis for the Zodiac, it is a physical relationship that can easily be observed and measured. We will see this more clearly when we observe this from the point of view of the Earth. If you can, follow along with your globe to help with the diagrams on the next page.

When we place the earth in the center we will have the tools we need to create Nikola Tesla's birth chart. And we will begin to show that the earth's tilt does more than make the seasons different it makes people different.

Spring Equinox
Beginning of **Aries**
(Approx. Mar. 22)
Equal 12 Hour Day and Night

North-Pole & South-Pole EQUAL

South Pole leads orbit

Fall Equinox
Beginning **Libra**
Equal 12 Hour Day and Night
(Approx. Sept. 22)

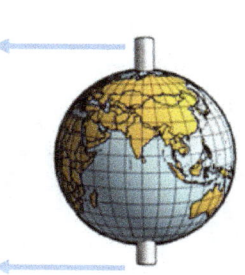

North-Pole & South-Pole EQUAL

North Pole leads orbit

Summer Solstice
Beginning **Cancer**
(Approx. June 22)
Northern Hemisphere – Summer
Southern Hemisphere - Winter
Longest day in North.

North-pole nearest

South Pole furthest

Winter Solstice
Beginning **Capricorn** (Approx. Dec. 22)
Northern Hemisphere - Winter
Southern Hemisphere - Summer
Shortest day in North.

North-pole furthest

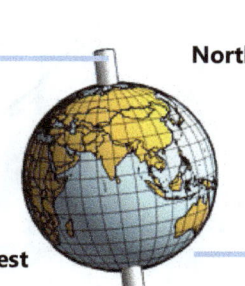

South Pole nearest

VIEWED FROM THE EARTH

Imagine everything revolved around the Earth. Below we see the spring equinox, the true Aries - the beginning of the Zodiac. Astrologers calibrate to this point to create the birth chart. Now it is easy to see where the 12 signs of 30 degrees each comes from. Again, there are absolutely no star alignments in the birth chart. Now it is becoming increasingly clear that the interaction of the planets is the true basis of the Zodiac.

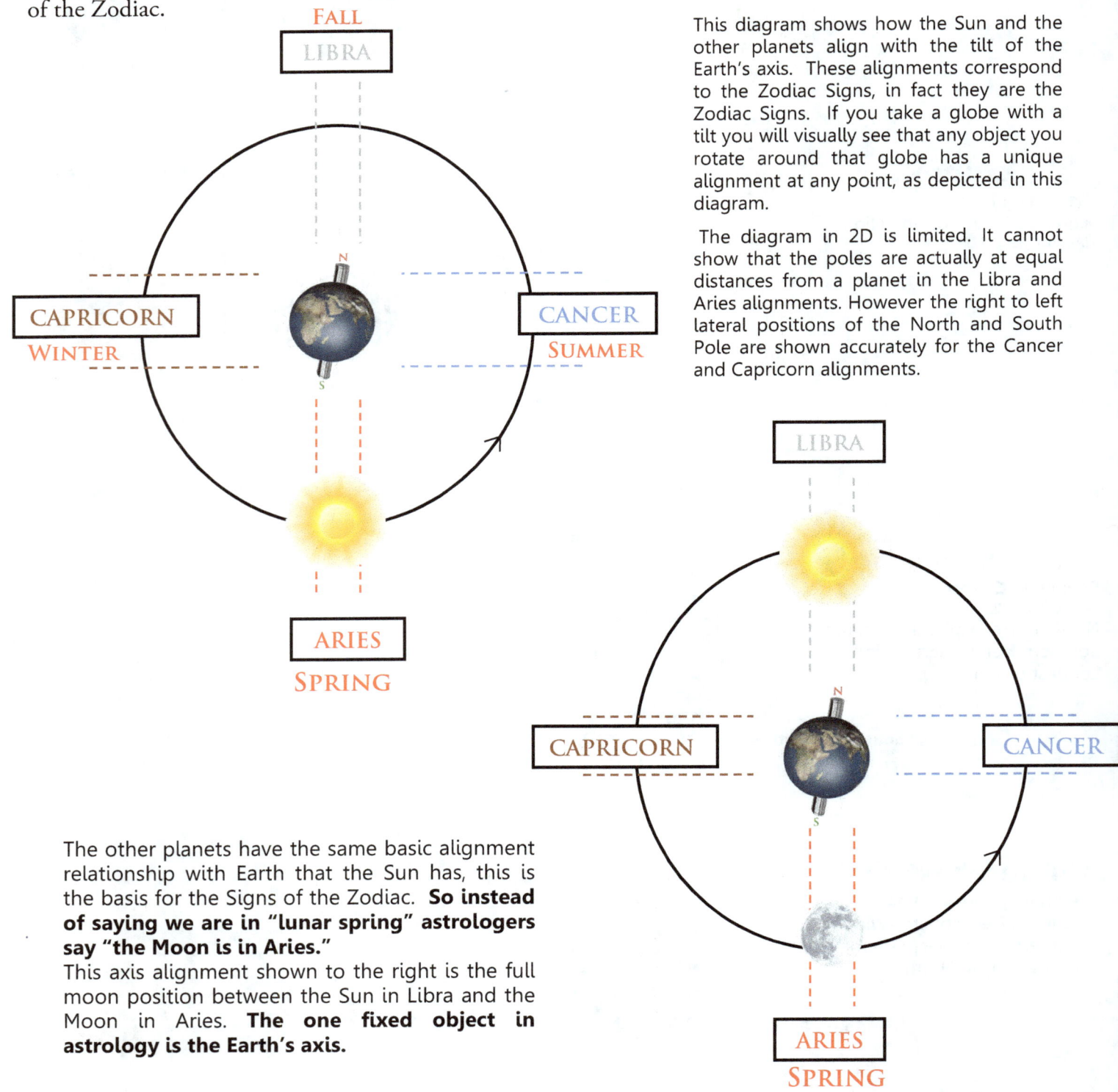

This diagram shows how the Sun and the other planets align with the tilt of the Earth's axis. These alignments correspond to the Zodiac Signs, in fact they are the Zodiac Signs. If you take a globe with a tilt you will visually see that any object you rotate around that globe has a unique alignment at any point, as depicted in this diagram.

The diagram in 2D is limited. It cannot show that the poles are actually at equal distances from a planet in the Libra and Aries alignments. However the right to left lateral positions of the North and South Pole are shown accurately for the Cancer and Capricorn alignments.

The other planets have the same basic alignment relationship with Earth that the Sun has, this is the basis for the Signs of the Zodiac. **So instead of saying we are in "lunar spring" astrologers say "the Moon is in Aries."**
This axis alignment shown to the right is the full moon position between the Sun in Libra and the Moon in Aries. **The one fixed object in astrology is the Earth's axis.**

The Earth and the Signs

The alignment of the Sun and Earth show the true origin of the 12 signs of the Zodiac as used in individual astrology. The 12 Signs relate very closely with the calendar year of 365 days divided into 12 months. Basic astronomy tells us that the planets orbits are not circles but ellipses, so the match isn't exact. However, 360 degrees is valid for the birth chart with each of the 12 signs being 30 degrees each; this is very accurate for charting planetary positions around a spherical object. The Earth's axis alignments are well known when the celestial body is the Sun, yet the only meaning commonly ascribed to them is obvious seasonal changes, but as has been shown there is much more to these alignments to be explored, both within astrology and science in general. These same alignments exist with the Moon and the other planets as shown with the Sun. The Earth's axis has been revealed to be the true center of the birth chart. Graphically you can basically take Earth centric diagram of the equinoxes and solstices and superimpose it directly onto the birth chart. The tilt of Earth's axis and its spin are critical because these are physical phenomena that we can scientifically evaluate. The descriptions of the signs start from the equinoxes and the solstices then these descriptions were given to star constellations. The zodiac signs never started with the stars.

With the 4 cardinal alignments sub divided into 3 equal parts each for a total of 12, we now have a clear physical correlation to the 12 Signs of the Zodiac, a configuration that requires no stars whatsoever, and more importantly a configuration that gives us a solid starting point for uncovering the science behind it.

Rotate the diagram 90 degrees clockwise and it matches the birth chart exactly.

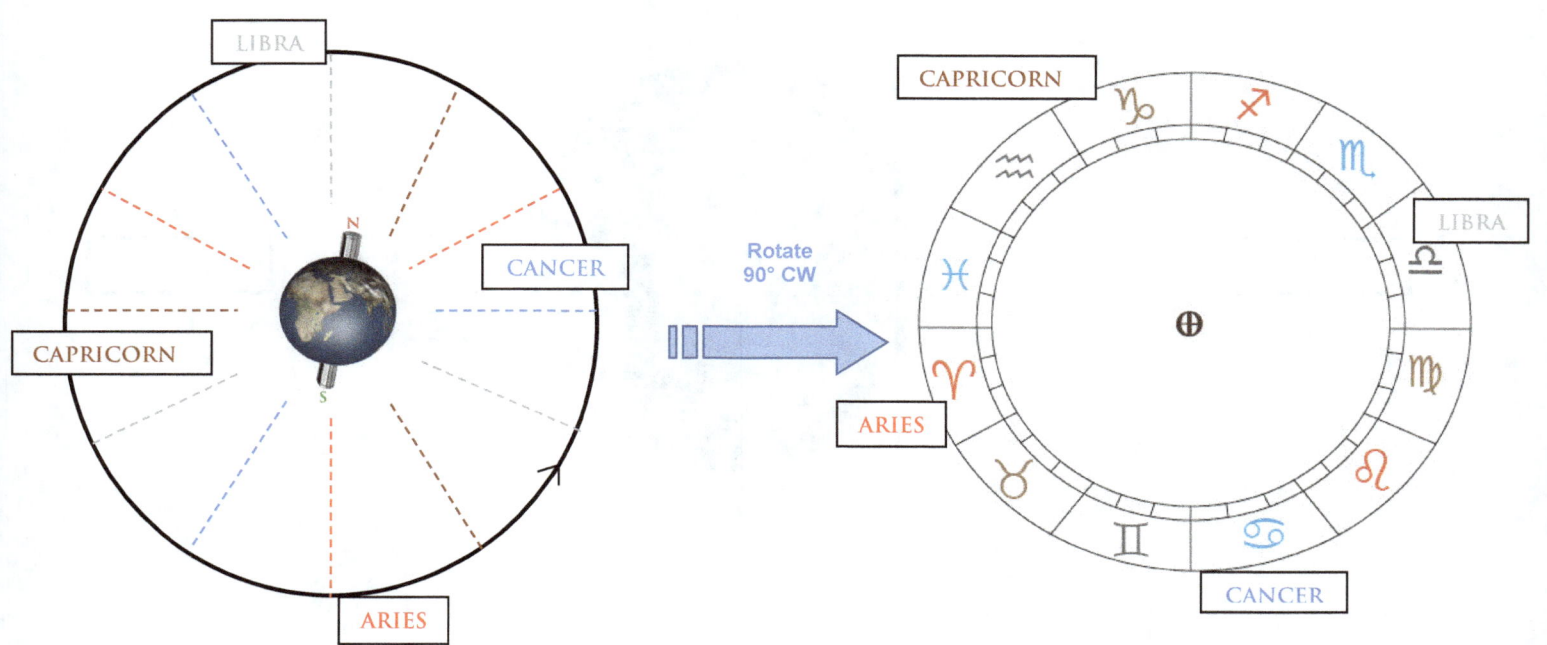

Adding The Earth's Spin

You know what your sun sign is, but do you know your rising sign? The rising sign, or first House, is often said to be the second most powerful influence in a birth chart. Rising sign is described as a person's outward persona, or the qualities that are encountered when initially meeting someone. The earth's spin is what gives us the 12 houses, the second part of the birth chart.

Let's build on what we know about the earth's axis producing the signs. You'll need a globe to follow along. *The globe is the key, keep the axis stationary and not only do the planets move but so does the earth's surface.* Below you can see, the earth, takes 24 hours to complete a spin, so there are 2 hours per House and sign change. The first House is 90 east of any position – eastern Horizon. In the below example the 1st House would be Aries, or the "rising sign" seen on the horizon at the exact moment of birth. Two hours later the rising sign would be Taurus. In fact, the earth's axis tilt is the only physical explanation for the total creation of the birth chart, the 12 houses, and the 12 zodiac signs.

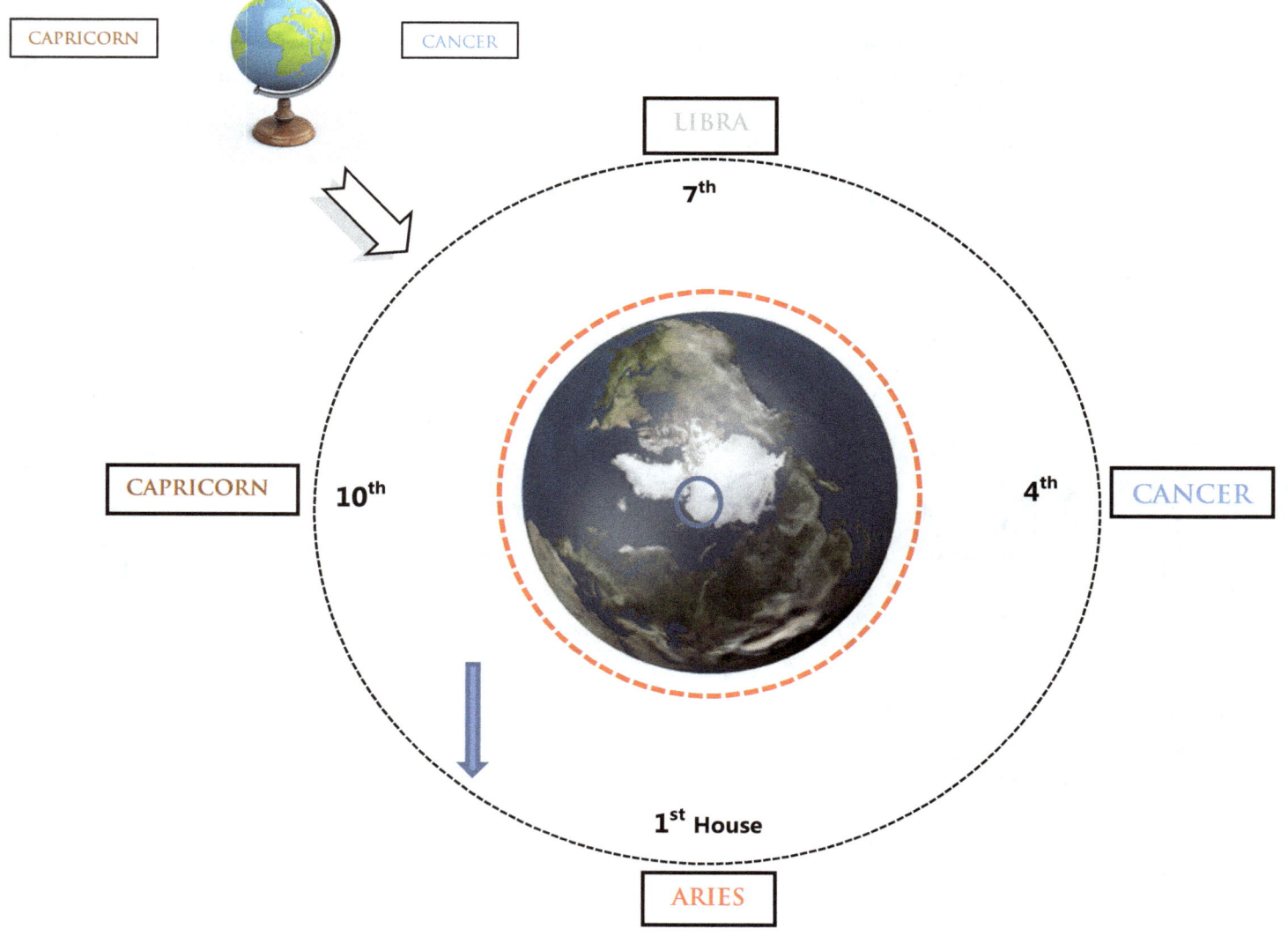

Representing The Houses

Basically "your sign" corresponds to the day in the year of your birth, and your Rising Sign and the Houses correspond to the exact time of day of your birth. Understanding signs and houses together may require a bit of effort at first so it's important to go step by step to make it all very clear.

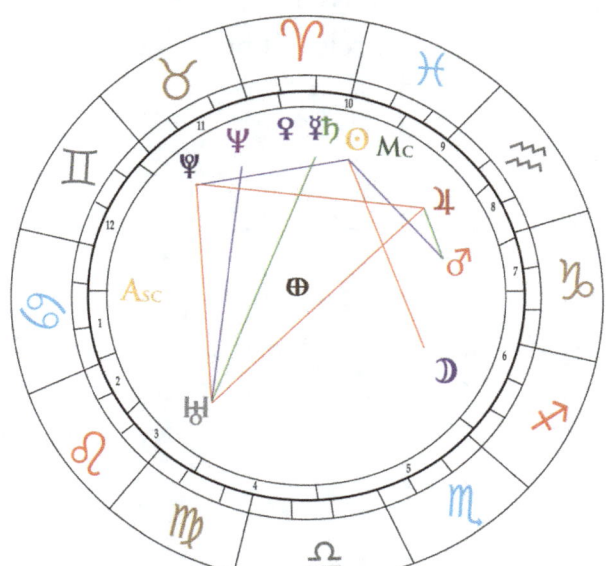

Resulting chart Cancer rising and Moon in Sag, Sun in Pisces.

The Houses also highlight dynamic relationships that are present in everyday life. These occur between the 1st house, which represents "The Self" (also called Rising Sign or Ascendant - ASC, that forms a natural opposite to the 7th House that represents Relationships, and the 4th House of the home and domestic issues is the natural opposite of the 10th house of Career. These are very common dynamics in everyone's life, the balance we seek between genuine self-interests and the desire for companionship and to be part of a union. There is also the drive for success in the outer world that can conflict with the need to fortify and maintain a home base and domestic life. The challenge is to find balance between these areas of life, but it often requires significant effort to achieve such a balance. It is much easier to become unbalanced, whether its losing one's self in a relationship or destroying relationships through selfishness and self-centeredness. Also sacrificing a career in order to raise children, or being so consumed by work that family bonds and domestic life suffers. It is the overall purpose of the Houses to show where natural inclinations exist and to highlight those areas for each individual. Pluto strongly placed in 10 house, common among presidents, Bill Clinton and JFK.

House Meanings

1st **House:** The Self
3rd **House:** Mind & Communication
5th **House:** Talents
7th **House:** Relationships
9th **House:** Expansion of the Mind
11th **House:** Friends & Groups

2nd **House:** Possessions
4th **House:** Home & Family
6th **House:** Health
8th **House:** Death & Mysticism
10th **House:** Career & Public Life
12th **House:** Self-Undoing

THE BIRTH CHART

Here we see the birth chart of Lewis Latimer, inventor of the lightbulb filament. Making up the outer ring are the twelve signs and the inner ring with numbers 1 to 12 are the 12 Houses (which can vary in size). Inside of that ring are all of the planet positions at the time of birth from the Sun to Pluto. All of the crisscrossed lines are significant angles the planets form with one another around the 360° circle. These angles ranging from 0° to 180° have powerful meanings within the chart.

Birth Chart
Basic Factors

✴ Planet Positions in Signs
✴ Planet to Planet Angles
✴ Planets in Houses
✴ Signs over Houses

☉ Sun
☽ Moon
☿ Mercury
♀ Venus
♂ Mars
♃ Jupiter
♄ Saturn
♅ Uranus
♆ Neptune
♇ Pluto

Lewis Latimer
Monday Sept 4, 1848
Chelsea, MA
Long:71°03W Lat:42°39N

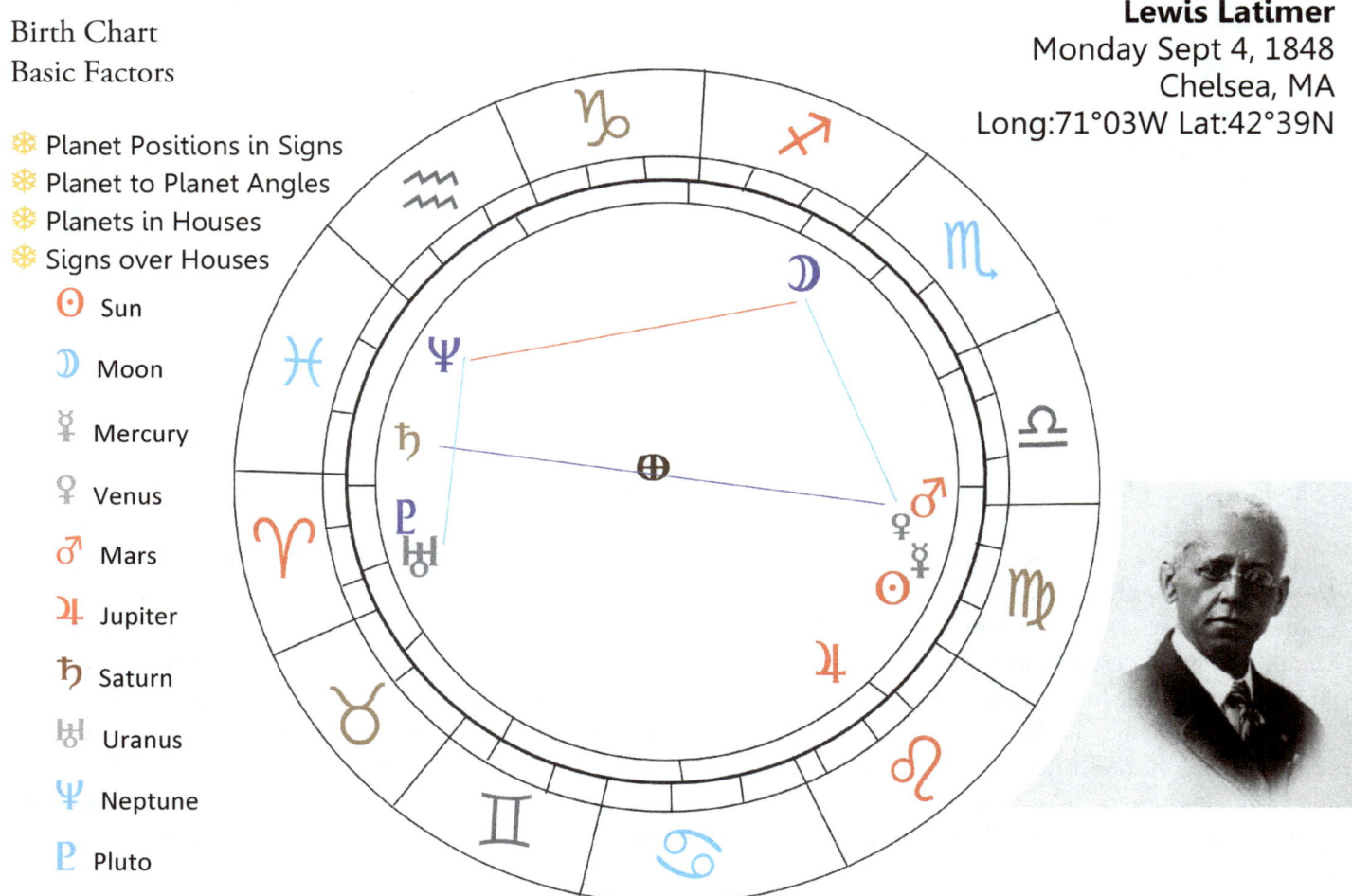

All of the basic factors are combined together to form a birth chart reading. The process of creating a reading is both an art and a science. The chart data alone is a matter of fact, but sometimes depending on the "interpreter" a given interpretation of that information may contain more or less insight. The process is not arbitrary each factor has firm meanings and principles, and it is the overall objective of chart "interpretation" to explain how all of these factors and co-factors combine in a unique way that is reflected in one's basic personality.

The Birth Chart

Below is the birth chart of Nikola Tesla. Tesla's chart shows a very powerful placement of Pluto directly on the Rising Sign. According to astrological interpretation, this is an extreme factor of influence for self-direction, inner-motivation and insight into the inner workings of "things". In Astrology there are other several powerful placements of planets, these are often found in the birth charts of many powerful figures, inventors, tycoons and moguls. The placement of so many planets in the lower half of the chart is also a classic sign of tendency toward introversion, or a distain for public life. In Einstein's birth chart there is a heavy emphasis on public life with a large degree of self-fulfillment and expression surrounding career and public standing. An astrologer would also point out a strong philosophical influence given the placement of the Sun in the 9th House and once again a Pluto-Jupiter combination indicating someone who goes against the established norms; in Einstein's case, challenging the norms of science in his day.

Nikola Tesla
Thurs, July 10, 1856
10:58pm Smiljan, Croatia
Long:15°19E Lat:44°35N

Making of the Birth Chart

It all starts with the solar system. The normal view of the solar system is with the Sun in the middle, as shown to the right. The rings show the basic order the planets orbits. The Earth is on the 3rd ring.

Next, the sun moon and other planets are moved to the outer edge but they keep their positions relative to the Earth's tilt. Notice, we aren't using any stars!

Finally, we get the birth chart. The planet are now shown as symbols with the Earth in the center. The large outer circle has the Zodiac symbols, which are axis alignments with Earth. The same that give us the solstices and the equinoxes.

Astrology's Time Cycles

The Yearly Cycle

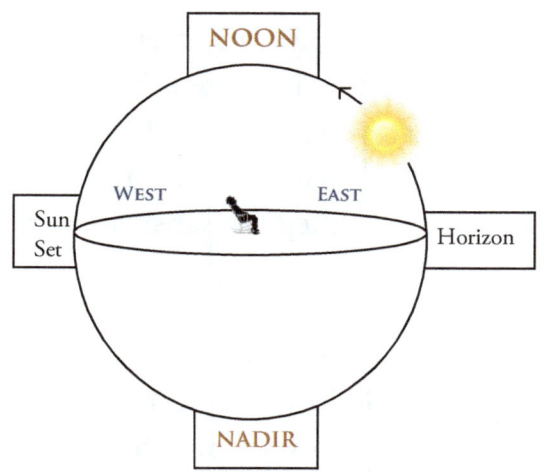

The Daily Cycle

The Ages

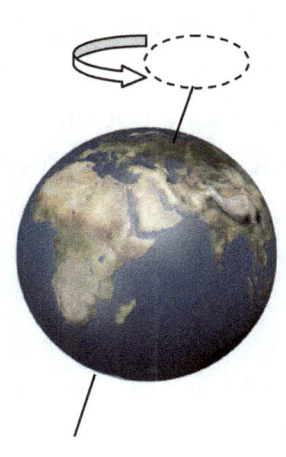

This is the cycle used to determine your Birth Sign or Sun Sign. These 4 cardinal alignments correspond to the actual beginning of the seasons on earth as well as the four cardinal signs of the zodiac, Aries, Cancer, Libra, and Capricorn. The Zodiac is then divided into 12 signs just as the calendar is divided into 12 months.

The Signs themselves are a result of specific alignments based on the earth's axis. The importance of the yearly cycle is that the sun sign is the strongest single influence and is calculated first. And is the basis for recording the position of the Moon and planets such as Mercury.

The spin of the Earth around its axis as viewed by an observer on the ground, the view you get if you stand outside in a clear landscape and observe the sun rise and set in the visible night sky looking in the sky from east to west.

In Astrology the exact time of day a person is born is as important as the Sun sign. The sign that is visible on the eastern horizon at the exact time of birth is called the Rising Sign. The alignment at the horizon also sets the first position of the 12 Houses of Astrology. However, it is the actual physical relationship between the Earth's surface and its axis tilt that is the true cause of this as we previously covered.

A celestial movement causes the earth's North Pole to slowly course a circle. The full circuit through the dotted circle pictured above takes 25,920 years this is called the great year.

The great year is divided into 12 ages, of 2160 years, using the same 12 Zodiac names used in the yearly cycle of Aries through Pisces. This movement is called precession, but unlike the yearly cycle which goes forward through the signs, precession goes backward. We are now coming to the end of the Pisces Age and entering the Age of Aquarius. This time cycle is not used in the birth chart.

*There is also progression and rectification which are beyond the scope of this book.
**There is a secondary wobble along with the axis cycle, but has no effect on the topics covered.

SECTION 2
THE DETRACTORS AND DEBUNKERS

Astrology's debunkers come in many varieties, ranging from scientist to opinionated skeptic. But the truly solid arguments are actually few, and the vast majority of the detractors fail on the first step. Most debunkers attempt to make astrology appear nonsensical by trying to expose what they assume is Astrology's basis; the star constellations of the Zodiac. We have already seen that these constellations are not used in astrology. In most cases, they are satisfied with this "expose'", and declare victory.

Unfortunately, many of these public figures have large followings and affect the thinking of broad audiences, resulting in widespread misinformation. Simply reading a competent book on Astrology that shows how to create a birth chart would have prevented the majority of these arguments.

Fortunately, in the previous section we cleared up many of the major misconceptions and explained some of the inner workings of astrology, the Earth's axis tilt and alignment of the planets. Unfortunately, we still have to deal with the misguided and uniformed arguments that are still rampant in the popular media. Testing accurately requires a firm grasp of the basic concepts of Astrology not just surface assumptions. It is no wonder why many popular debunkers have so many self-proclaimed victories.

Many people know there is some truth to astrology, enough truth to at least not dismiss it blindly. But some scientists, with their narrow approach and misinformed arguments would have us believe we are under mass delusion. Almost as if they are the sole authority of what is fact, and if they decide the sky is green, then we all must be color blind if we see blue. When confusion is allowed to persist, misinformation and falsehoods prevail. It's time to end the confusion.

Arguments Against Astrology

We have already established the true zodiac, shown the birth chart, and that astrology is a solar system science closely related to the seasons. So it almost seems redundant to deal with the question of the stars again. But it is still important to explain why this business of a "13th Sign" doesn't matter. First of all, in the space making up the zodiac ring around the Earth and solar system there are 14 "established" constellations not just 12.

How many of the detractors points can we address? Let's take the top three.

1	How do star constellations Affect us?	☐	There is no Claim that they do.
2	Astrology has no physical basis	☐	Yes, it Does. The alignment of the earth's Axis with The sun, moon and other planets.
3	How can planets Affect Us?	☐	There are possible ways within the current understanding of science. Part 2 explains this. There are also related effects Accepted in mainstream Science.

In any case for the uninformed and the misinformed it turns out that stars are not used and have not been used in personal astrology. Also common sense tells us closer and more massive objects and forces tend to be stronger, an intuitive observation that turns out to be correct. This is exactly what we find in the birth chart which is based on our local solar system.

CRUSH WEAK ARGUMENTS

The star constellations change due to precession, a fact known by astrologers and detractors. The precession argument is easily defeated by bringing up the birth chart, and its use in astrology. Then the fact that astrology is based on alignments between the sun and the planets in our solar system not the far away stars, ends that argument completely.

Astrologers are well aware of the constellations moving and use this precession to keep time with the ages of the zodiac which cover large thousands of years at a time. The ages are not used in individual astrology, which uses shorter time frames of years, days even hours.

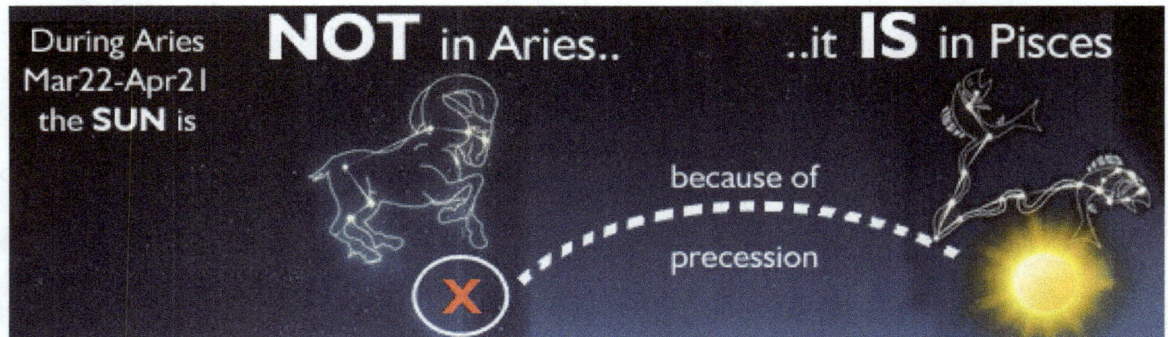

The birth chart is the first document produced in astrology. The second is the horoscope, which is based on the birth chart. Sun sign horoscopes are based on a fraction of the chart, a small percent of the factors in the birth chart. The funny thing is that the Sun sign is the second most challenged part of astrology. Once again bringing up the birth chart and all of the planets in it will end that argument. And explain why any tests with just Sun signs will fail miserably. Any test of astrology must test the entire birth chart.

Why is the birth chart never brought up by these debunkers? The birth chart can only be missed if you were not using science to analyze the claims of astrology in the first place. The scientists that refuse to use science to analyze astrology is a long list that includes some big names like, Neil De Grasse Tyson, and Bill Nye. Ironically, Neil Tyson himself said, "... scientists don't always use science." Sadly, he and his colleagues have not used diligence to truly investigate astrological claims, otherwise they would have found tremendous repeatable evidence in the birth chart - and a great mystery to be solved.

Testing Astrology

There are a number of "experiments and tests" of astrology on the internet even in college physics and biology lab books. The sun sign is the usual focus of these experiments, however the sun sign is only one of many astrological factors that influence the entire individual personality, the entire birth chart is needed if you want specific and useful information. But virtually all of these experiments show absolutely no knowledge of the birth chart. There is actually an experiment of listing the entire classes' signs (not even birthdays) and their majors and seeing if the signs and majors match, under the premise that astrology should show career aspiration. This is like trying to test drive a car without knowing what the gas and brake pedals are for, or trying to experiment with magnetism using two pieces of plastic. Where does astrology claim that sun signs dictate professions? These experiments are just silly and ridiculous, and there is never a mention of the birth chart, which does contain those kinds of details that can indicate affinity in certain professions. These are sadly uniformed experiments lacking the basic knowledge contained in the average astrology book.

Good tests require a valid basis to start from. Deduction requires isolation to be able to test given factors, such as the sun sign. The problem is that the sun sign is all they know, they have no way to isolate it without the birth chart. The solution is to read and test for yourself. This book will only cover the basic principles of astrology. To become properly informed on astrology as it is practiced, there are easy to read books that cover the essential topics. A list of these books is at the end of the next section.

And for those that want to truly test astrology. There is an excellent test. Create your birth chart and the charts of two people you know. Then do one of someone you barely know. Do the compatibility between you and the two you know. Then see if you still think astrology is complete nonsense. Challenges are welcomed. When something is true, it can always be questioned. But, to know what questions to ask, you must be properly informed.

Most of the people who are devoted to astrology are not devoted because they simply believe everything they read in books, it is because they have discovered truths in everyday life. They are not hindered by the imperfections of astrological narratives. The focus becomes the truths that they have been able to see clearly.

SECTION 3
Astrology In Everyday Life

Astrology is one of the few subjects that can be completely about you. Will you be rich and famous? Who will you marry and when? In this section, the basic principles of astrology according to astrology will be covered in a concise way to tie in the physical basis we have already covered with the deeper meanings that astrology is more commonly known for. Astrology can be a very useful guide for self-discovery and self-knowledge. Once the basic understanding of what the birth chart represents is achieved, the knowledge gained can be very useful in understanding personal issues in everyday life.

Another true value of Astrology, is detailed in the principles of signs and planets and how all of these factors interplay in the birth chart to give a very accurate and specific description of an individual. The knowledge of the planets and signs seem to have been acquired and tested over thousands of years of observation and correlation. A period of validation that is almost unique in any formal practice, besides the practice of astronomy. This is extremely specific information describing an individual's personality traits. Taking a close look at the specific factors in the birth chart; Planets in Signs, Planets in Houses, and Planet to Planet angles called Aspects, for an individual at birth reveals a very clear connection to personality traits that are hard to dismiss as coincidence.

Astrology claims several basic things; a profound influence on personality, the ability to forecast when life events may take place – prediction, compatibility between individuals, and a lastly it lays out a "System of Order" connecting all the experiences in our lives to the motions and cycles of the planets.

SIGNS & PLANETS

A good way to think of each sign is as a distinct tune or vibration and the planets as instruments. The signs are organized into 4 harmonious groups by element. Each planet is also associated with an element. Generally "empty" signs in the birth chart have very little if any effect. For example, if there are no planets in Scorpio in a particular birth chart then Scorpio will be "muted" (although it may play a role in the Houses), and the individual will not show those traits. No planet, no expression. A tune cannot be played without an instrument.

FIRE – ♈Aries ♌Leo ♐Sagittarius
Dynamic, Creative, Self-Expressive

EARTH – ♉Taurus ♍Virgo ♑Capricorn
Practical, Methodical, Grounded

AIR – ♊Gemini ♎Libra ♒Aquarius
Outgoing, Cerebral, Excitable

WATER – ♋Cancer ♏Scorpio ♓Pisces
Feeling, Nurturing, Intuitive

Listed above are the basic principles (the positive aspects) of the elements. Each person has varying amounts of these elements and principles uniquely expressed in our individual personalities. If you see an attribute listed above and it describes you, but your Sun sign is from a different group, that's fine. Remember, astrology is based on the birth chart which means each person is essentially a combination of all the elements in varying amounts.

Fire and air are natural compliments. Fire needs air to ignite and continue burning and air is in turn stimulated by fire. Water and earth are natural compliments. Fertile ground produces life when watered, and shaped earth can provide direction and base for water. From the basic principles we get the range of personality types and the rules for compatibility between these types. The signs create the chemistry and compatibility, we all know are real.

This is not something that predicts the future. But when you know how the signs work it is a guide, like a having a flashlight in a dark room. The value of the birth chart is in showing exactly where these qualities are concentrated and how they are expressed specific to each person. So having your Sun in a certain sign and element only means that those principles should be strong. But there are still many remaining factors involving the other planets and which signs and elements they are in.

Signs & Planets

In astrology, the planets are the key to an individual's personality. The most important factor is the placement of the planets. Which signs the planets are positioned in is what determines which signs get expressed. The planets themselves are associated with elements as well. When a planet is "in" a particular sign, it takes on the characteristic of that sign within the life or personality area that that planet specializes in. For example, Mars in Gemini would combine the aggressiveness of Mars with mental talents of Gemini. This could result in a strong intellect or someone who starts debates even arguments.

The second most important factor is the angles the planets form with each other. A common example is when two planets are 180 degrees from each other. In Nikola Tesla's chart, the Moon is opposite Jupiter. This creates emotional excess, and the possibility of misplaced emotional belief in charity with very little regard for real practical gain (among other possibilities).

Another factor to look for is the harmony in the chart itself. The inner harmony of the individual person. This principle of harmony is also used when finding the compatibility between two people. It's the job of the astrologer to take into account each separate detail and combine them together to provide insight to assist with everyday life. Each chart is different, so it can be very challenging.

☉ Sun - Self-Expression

☽ Moon – Emotive

☿ Mercury – Mind & Communication

♀ Venus – Beauty & Romance

♂ Mars – The Physical Self

♃ Jupiter - Expansion

♄ Saturn - Discipline

♅ Uranus – Mental Depth

♆ Neptune - Imagination

♇ Pluto - Power

THE PLANETS INTERACTING

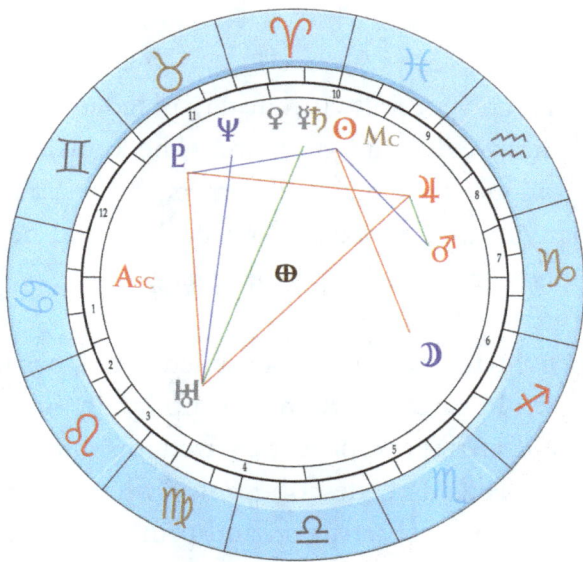

In the example chart pictured above, we see the tremendous power found in the interaction of the planets, by the angles formed between them. These special angles, called Aspects, help to explain some of our more detailed personality traits and show the inner workings behind how some of our traits are more intense than others. These special angles also show how the overall mixture of traits are a blend of some complementing traits while others are conflicting and give rise to our inner conflicts.

0° Conjunction - Very Strong combination can indicate uncommon abilities, but can be altered by another other planets.
120° Trine - Extremely Favorable

60° Sextile - A complement, a general positive combination.

180° Opposition - Can be challenging and limiting, but depends heavily on other planets being involved, which could have a positive or negative additional effect.

90° Square - indicates challenge, difficultly, conflict, limitations and possible excess.

In the example, the Sun square the Moon indicates the inner and outer self in conflict, resulting in difficulty in emotional expression. It can be very helpful to highlight these issues and begin to understand and manage them. But Mars makes a favorable 60 degree angle with the Sun. The planet Mars defines physicality even down to sexual tastes, preferences and performance. The chart is a complex mix of factors. A positive angle to a planet may accentuate, where a negative or challenging angle may cause perversions or severe sexual inhibitions. It's the job of a good astrologer to list each one of these angles into an Aspect List and rank them by strength.

The Individual

One of the key uses in the ancient and current day practice of Astrology is to provide the individual the advantage of self-knowledge and guidance. The knowledge gained provided by the birth chart can be instrumental in avoiding major life errors and assist in the self-management of basic tendencies. Although many of us are aware of our positive and negative tendencies at some point in life, it is often useful to get an objective perspective.

The birth chart is also used to indicate innate talents and naturally strong abilities that people seem to be born with, even in cases where there is no family or outside influence. In total the combined information should attributes, potential strengths and faults all of the makeup of the personality. The description is said to describe only "what is already there or potentially there", because Astrology is a practice of observation and principles taken directly from the birth chart gained from thousands of years of human observation. Can help you know yourself as well as others.

In astrology the angles formed by the position of the planets are called aspects. Each aspect shows a list of attributes and tendencies, but also potential paths for self-development. Planetary positions grouped in certain houses show strong emphasis in certain areas of life. When this is combined with the factor of planets in signs, then the angles between the planets show compliments or difficulties in those areas of life. Signs aspects, houses, planets, combine to yield a vast matrix of highly specific traits, from narcissism to emotional dispositions, even sexual tastes.

The insight and increased knowledge of yourself can also assist in being able to know others more deeply and meaningfully as well. The use of the word "personality" refers to a collection of traits that influences all areas of life.

Many areas of life are dictated by our responses to the environment. Even traits to reject the indoctrinations of society are described by the birth chart. An individual's focus in certain areas of life is heavily indicated by the Houses, and the planets inside of them. The question becomes, is your life more a result of your directed action or the sum of circumstances outside of your control? The answer can only really be addressed on a case by case basis. But it's an excellent question that requires more than what the birth chart can offer alone. Fortunately, it is a question that is the subject of Part Two. But many other topics must be covered and assembled before we can properly address the important question concerning the factors that determine our lives.

THE POPULAR HOROSCOPE

The well-known horoscope of astrology is mostly known for being the general daily predictions for your particular sun sign. The obvious problem is, how can every Virgo, or whatever your signs is, have the exact same horoscope? Any decent astrology book will tell you that a true horoscope uses your entire birth information.

The proper creation of a horoscope incorporates the entire birth chart, the snapshot of planetary positions and your exact place of birth. A horoscope is your planetary positions at birth compared to the current position of the planets. This produces a reading and forecast of influences that can last for short and long periods of time. These professional level horoscopes can be made with expensive software systems that automate the basic interpretations of these influences.

When done properly a horoscope can forecast life events, certain auspicious times, like the potential meeting of a new love interest, or a specific time period that may benefit career advancement or financial gain. Of course, there is a great deal of controversy surrounding prediction. Is it really a prediction of events that would happen if you did nothing, or is there an influence given by the planets that influence you to think and act in a way that makes certain events more likely?

Ultimately, you will have to be the judge of the validity of any influences and interpretations that may be in a true horoscope. Although horoscopes can provide great insight they must be viewed with caution and as potential influences, not as predictive. Just like any forecast, they are meant to highlight auspicious times and possible circumstances that may arise. Even accurate horoscopes based on birth charts cannot account for all factors, a larger system is needed that will be introduced in part two.

In my experience, the predictive claim of astrology is not very accurate. Too many things have to come together for some predictions to take place. There are times when everything comes together and the influences of planets may play a larger role, but they do not determine anything by themselves. I will say, major events in horoscopes tend to be more accurate than daily or minor events. The experience of doing many horoscopes for many years has proven that there are just too many cross factors to make it like a script for your life. It simply is not.

Still Subjective

The analysis of the birth chart can have some strong conclusions, but even then there can be some subjective selection of what is influencing which area of life or behavior. Not only within astrology, but there is also a valid argument that some of the influences on the individual will come from experience and environment. Astrology has not developed a method to deal with this reality.

Taking a comprehensive view, it is still apparent that the birth chart has a significant influence on the individual, but it must be accurately represented as a part of influence in a larger context that includes all factors affecting individual traits and development. So many factors for a given personality trait, like musical creativity; sign of Pisces, planet Neptune, aspects with Neptune, too many to list here.

Another example is an interest in medicine or the medical field. This has several different possible sources of influence. Most of the influence surrounding medicine come from the sign of Virgo, the planet Mercury, and the 6th House.

In astrology there are many personality traits that have many sources within in astrology alone. It takes a good deal of time and practice to be able to combine and resolve all of the factors of influence in the birth chart. For some, the birth chart may be interesting, but not convincing. Someone may try to point out inconsistencies or explain away a dead-on description of their personality as coincidence. But the next level of proof is much harder to argue against, and that is astrology's system of compatibility.

Synastry in Astrology

Synastry, is a word that is the combination of synergy and "chemistry"(the kind of chemistry we experience within relationships). Synastry, as used in astrology, is simply combining 2 birth charts to see how two individuals relate to each other, to determine how compatible or incompatible they are.

The fact is, that the basis of human compatibility is far from subjective, it can be highly objective as will be demonstrated in the coming sections. It is worth pointing out that early Astrology was simple observation the ancient societies used the stars and "celestial bodies" as a referent for as many events on Earth as they could correlate. The attributes of human personality and especially human compatibility were observed to correlate to planetary positions, this started of course with the sun and the moon. The point being, there was nothing to "make up" it was and is not a form of mythology. The frequency of these planetary combinations amongst relationship partners is extremely high. More importantly the aspects in combination give detailed information on the inner workings of many of these relationships.

Classic Planet Combinations of Astrology

☉ **Sun &** ☽ **Moon** – tolerance, friendship and broad commonality

☉ **Sun &** ♀ **Venus** – romance, sexual rapport, appearance attraction

♀ **Venus &** ♂ **Mars** – sexual rapport, appearance attraction

The above are just some of the most common combinations of planets that are classic signs of compatibility, when the planets of two birth charts are compared. For example the Sun of one partner in conjunction (0° angle) with the Venus of the other partner in the sign of Cancer would produce an undeniable chemistry between a proper match of appropriate ages. We can self-analyze sitting in a room all day, but we really come to know ourselves through our relationships with others. And your reaction to other people is as real as it gets.

Synastry in Astrology

The one thing that we know absolutely is our like and dislike for others, whether it's right away in an initial meeting or whether it takes time to confirm. We simply know who we like and who we don't. No one has to convince us of this. This "power of recognition" is reflected very clearly in great detail within the comparison of two individual's birth charts. This amazing natural ability we all have to know our own reaction to others called compatibility recognition, we use it in all types of relationships, whether it's a friendship or a romance, even within our own families. We can do this even beyond initial superficial signals.

An incredible feature of astrology is the ability to point out the details of "what is already there" with regard to compatibility and incompatibility. Common experience like; "We just clicked", "there's just something about her/him I just don't like", "We have a good vibe" the list goes on and on. Learning how to distinguish between merely being physically attracted to someone and recognizing deeper connections that take time to confirm. Even though you have to confirm it, it was "still there" even before you met, it just took the meeting for it to be realized.

The convincing thing is when these reactions are strongest the comparison / compatibility charts of astrology indicate these reactions very clearly. However, they do not totally dictate the outcome of such relationships, the charts show potential possibilities.

There is a fundamental reason why certain musical notes harmonize, why certain colors match and others don't match. This same principle is present in human interaction and influences much of our relationships and compatibility. It's the chemistry you feel when you first meet someone who you feel like you've always known. In a roundabout way in order for this compatibility system to work it must have been valid in the individual case in the first place. Beyond the individual there are the personality types. The 3 step process; Cross comparison & compatibility comparison.

Within the basic make-up of each individual there are specific "elements" of personality that form compatibilities between individuals. It is at this level that the subjectivity that can sometimes cloud the interpretation of individual birth charts is discarded. The planets and factors of compatibility are clearly identified.

Astrology can show attraction, chemistry, rapport and areas of compatibility and conflict. And can help confirm some of the more difficult initial unknowns. But there are still limits, astrology cannot tell you if a relationship will work, you still have to make good choices. No other field of study can even come close. DNA will not tell you anything about personality, without seeing two people. Any doubter please test the birth charts and see for yourself.

You cannot deny your instincts; your feelings, your attraction, your indifference or your disdain for other people. You simply know it. Somehow it is all part of our senses, as real as touch or taste. This is no accident, chemistry is shown by comparing two birth charts. In astrology, Venus is generally the central planet that combines with other planets to produce most of the chemistry we experience in our romantic lives. This can be analyzed with great accuracy by a third party with absolutely no interaction with the two individuals. No area of mainstream science can claim to even come close to this. And by knowing and understanding our traits and tendencies and the traits of others we can make better life decisions.

A simple test of going to a crowded place like a mall you can easily see people who you are attracted to, people who don't get your attention and possibly people you wish you didn't see. Attraction is as real a "measurement" as measuring the distance from your head to your toes. The difference is that you are the instrument used to measure what is attractive. Astrology is well aware of this and the mechanism behind it. Not only what is attractive, but who is generally attractive and when there is mutual attraction - what we experience as chemistry. According to astrology the planet combinations of mars and mars, mars and Venus, sun and mars and sun and Venus are at work.

Chemistry is not just physical attraction. Astrology shows compatibility across broad areas of traits. For example, we also experience connections with other people through conversation. If you talk about the right topics with someone you quickly get a sense of your thinking being in tune with theirs, or not. We all have friends we can talk to for hours with no effort. And then there are people we barely have a few words to say to. Again, astrology explains this type of communication rapport or lack of rapport with the planet combinations of; mercury and mercury, mercury and sun, and rising sign and mercury. The same principle works with moods being in tune. Astrology uses the interaction between the Moon and Sun, Moon and Moon and Moon and rising sign to describe this connection.

Another test, but less simple, is to look at the charts of yourself and your romantic partner(s). You can do this yourself or have a good astrologer do it for you. Chances are extremely high that what you experience in terms of attraction and chemistry are already spelled out for you in both of your birth charts – I know this first hand as a skeptic I have tried this many times. The results can be pretty amazing in the way that they detail both good and bad points in relationship areas, whether the relationships is current, past or never even happened yet.

Astrology's true usefulness is the insight it provides pertaining to the dynamics of relationships, how to understand the combinations of both partner's tendencies, not the "mythological" power of prediction. Also force of will and personality as well as certain circumstances can make people enter and stay in relationships with people they are not compatible with. The process of determining broad based compatibility is often left out of the selection process, in favor of lust, superficial attraction, infatuation and haste to be in a relationship and many other preexisting motives before the partners even meet. Taking time to evaluate one another to determine compatibility as foundational can help future success, long term happiness, and more harmonious relationships without harsh compromises.

Synastry in Astrology

If sceptics want somewhere to start testing, this is it. Dealing with qualities is different than dealing with quantities, analyzing quantities requires knowledge and analytical skills were qualities require wisdom and knowledge. Behavioral Science claims to be able to deal with human qualities, however their efforts are incomplete and limited without the knowledge of natural intrinsic human qualities.

Circumstances are often a more important factor in relationships at any stage initial or otherwise, and compatibility is a cofactor. The circumstance of receptivity is a key prerequisite. At any time there may be powerful counter circumstances, hidden and otherwise. For instance, your partner lied and was actually married, preexisting emotional connection to former partners, etcetera. Anyone can test these interactions out, all you need is the birth information. And even though astrology may be able to point out rapport, relationships are only part rapport, another important area is "effort and work". And it is always up to the two people to do the necessary work, and a good foundational rapport can make that work easier and more rewarding.

There are real everyday factors that prevent relationships from happening or being successful such as poverty and class.

This is still not a simple thing, you can test two charts using a book or books for reference but it takes experience and some insight to become truly adept at it, just like any other skill or field of study. Even more advanced is the threefold analysis, the two people separately and then the combination of charts to see the interactions.

The case of the man or woman who is "always" attracted to the person that is no good for them or not really compatible. Can be from an individual tendency also the fact that attraction is separate from personality compatibility. So even if the sun signs are traditionally compatible that does not mean there are romantic elements or physical attractions elements. There are seven main areas of relationships. Every match is a varied combination of these areas often there is simply no strong connection in any area, the chart will show that as well.

Just as with the individual attributes shown in the birth chart are not deterministic, meaning they do not fully control because you have choice. You could choose to ignore your tendencies and attributes and attempt to act in alternate ways but it would be difficult. Acknowledging your true strengths and faults would be more beneficial. It is the same romantic relationships, we often choose to ignore lack of commonality in favor of surface attraction only to find relationships unfulfilling and difficult. All of this without any analysis of birth charts where compatibilities and incompatibilities become very apparent. But the use of birth charts in the early going can help to reward your time and effort, even help you understand yourself and the other better, like a mirror for the relationship.

Recommended Reading:

For Overview and easy reading: The Only Astrology Book You'll Ever Need, Joanna Woolfolk

For Accurate Planet Aspects Interpretations & Chart making: Compendium of Astrology, Lineman & Popelka

For Romantic Relationships: Synastry: Understanding Human Relationships through Astrology, Davidson

SECTION 4
ORIGINS OF ASTROLOGY

In 2,650 BCE, on a hot June day along the banks of the Nile, a child was born in Africa. That child would grow up to become the first multi-genius in recorded history, today known as Imhotep. As a young child, Imhotep walked among great monuments, one of them Hermaket or Mehit, today called the Sphinx. One spring equinox morning, three months before his own birthday, while walking behind the great Sphinx he noticed the sun rising in perfect alignment with the center of the Sphinx's head. Like clockwork, he would see the same year after year. Today, we can also see the same. Yet the fantastic Sphinx is only one of many very old equinox markers. It curiously provokes the question, "what is so important about the spring equinox?"

And what do these equinox markers tell us about astrology's origin? These markers are a door to an ancient vault, inside is a powerful science – and we have unlocked it. The key is realizing the equinox is in both the monuments and birth chart. Then it all makes sense. The alignment markers are not only about the monuments, the alignments are about us!

The remaining stone pillars of Nabta Playa, once part of a stone circle, oldest and most sophisticated of all known ancient observatories. The site is located in the Nubian Desert (modern Egypt), and has been reconstructed by researcher Dr. Thomas G. Brophy in his book, 'The Origin Map'. Incredibly, the site is also able to show the accurate distances of certain stars, as well as precise measurements of precession and the timing of the spring equinox.

Just as the spring equinox serves as the beginning of the zodiac in the birth chart, we see the same alignment permanently encoded into ancient monuments. Knowing what we know about astrology, these incredible monuments are telling us, like voices echoing from the past, "yes we knew of our connection to the sun and the planets, we knew the science." One incredible example of this is an ancient site called Nabta Playa, a site dated at over 5,000 years BCE. The site doesn't directly show the practice of astrology, but it does show accurate measurements of the spring equinox and precession.

Origins of Astrology

There could be systems of astrology much older than even 10,000 years BCE, we just don't know. Given all of the known artifacts we simply do not have an exact date or time period. The reason for this is our new discovery of older sites and confirmed older dates for ancient monuments like the Great Sphinx, which may mark the constellation of Leo at the spring equinox at 10,000 BCE clearly demonstrating advanced astronomy and established constellations well before even 4,000 BCE.

And the many other ancient monuments mark several constellation alignments, as beyond last Aries and Taurus ages. The fact is we don't know the exact origin of either Astrology or the zodiac star constellations. All of these estimates are guess work. One thing we should make clear, is that solar and lunar calendars had to be created first, then the constellations were "created" and paired with certain times of the year.

The first step in any system of astrology is to create or use a calendar, usually based on the two largest and most obvious objects in the sky, the sun and the moon. The next step is to observe and record human behavior over the calendar year. The religious and practical aspects of observing the sky were likely practiced side by side. Worshiping the moon and sun as deities, measuring the year, predicting solar and lunar eclipses were seen as related in ancient societies.

Above a relief from the stone ceiling in Dandera (ancient Kemet - modern Egypt), shows all of the zodiac constellations within a calendar that our modern calendar comes from.

The Aztec Calendar in stone, identical to the 365 and 260 day Mayan Calendars.

At some time within the ancient practice of astronomy the idea or fact that people born at certain times of the year had different personalities was discovered. Just as certain times of the year required planting seeds while others times meant harvesting crops. So simply through observation alone systems of astrology could be created, just like systems of planting crops, hunting and migration also benefit from the invention of calendars.

Yet mysteriously, out of ancient history we have the appearance of the four oldest known systems of astrology; Nile Valley, Indus or Vedic, Sumer, and Chinese. Popular claims place Astrology's origin around 4000 BCE and attributing the origin to the last Aries age, circa 2000 BCE, is nonsensical. But as we have seen there are sites demonstrating related astronomical knowledge of the zodiac constellations thousands of years before that. In the Nile Valley's ancient Kemet not only did they have the same signs we have today, but they also used the signs to identify the longer 26,000 year cycle.

Astrology continued through the Greco Roman period and was practiced heavily in middle eastern societies. Moorish and later European astrologers would spread the practice further into Europe during the middle ages. Although we don't have any names recorded to give direct credit for the ancient systems, in modern times we certainly have several. Probably the most famous is Johannes Kepler. Kepler devised equations of planetary motion that gives a planet's position with having to see it with a telescope. And our recent advancements in computers have resulted in automating the calculations that used to be done by hand. It's also important to note that even though Kepler was a man of science, he also practiced astrology.

It is completely possible that our current astrological knowledge is only fragment and may be missing certain key truths. This is evident in the confusion of the stellar zodiac and the true effect of the Earth's axis alignments as source of astrology's signs. Fortunately for us, the entire process behind astrology is a natural occurrence that we can easily observe and put together the "missing pieces" for ourselves. A good place to begin searching for more answers are the various systems of astrology that have been practiced all over the world since ancient times. Could there be some clues or even a common thread between these various systems?

DEEP ANCIENT Pre 3000 BCE	ANCIENT After 3000BCE	RECENT & MODERN
(lost historical Ages)	Sumer Babylon	Moorish / Arabic 600s-1400s CE
Kush Nubia Kemet	Olmec Mayan	Renaissance Europe
Global?	China India Vedic	Western Modern
Gunang Padung	Greece Rome	Eastern Modern
Gobekli Tepe		

Zodiac Wheel-Middle Ages Europe

One clear common thread between all of the known systems is that they are all basically calendars. And no system of astrology uses stars to determine which sign you are. And lastly, all ancient forms of astrology come from societies that saw principles in all aspects of nature; elements like water and air, and various animals.

Chinese Astrology

The Chinese Zodiac is based on a 12 year cycle, with each year representing one of 12 different signs. The 12 Chinese signs are basically time periods and these time periods are based on a lunar calendar rather than a solar calendar. Each sign has its own unique set of traits and qualities, it is a broader system that groups people by year. However it uses almost exactly the same harmonic relationships seen in "western" Astrology. Compatible signs are grouped the same way, every 4 signs or every 4 years in the Chinese case. But opposite signs are viewed more negatively in Chinese astrology. The Chinese system can complement solar astrology, giving the broad stokes to the more detailed view of solar based astrology that is popular in the west.

The planetary movements seem to hold the key to a connection. Many planetary cycles occur in 2, 4, and 12 year increments. Venus and Saturn, two planets that are critical for compatibility, and Jupiter have movement cycles that seem to repeat within these time increments. This matches the planetary effect in western astrology. Not to mention the outer planets positions, which take years to move, can define larger groups' relationship to other groups born in distant time frames. This could certainly produce the personality harmony claimed in Chinese astrology. The connection seems to be the planets, the cycles of time and the relationships of how certain periods of time represented by signs are naturally harmonious with other periods of time. Check your birth year, you may notice compatibility patterns with others.

Chinese Zodiac Years – Compatible Groups
Rat 1972,1984,1996,2008
Dragon 1976,1988,2000,2012
Monkey 1980,1992,2004,2016

OX 1973,1985,1997,2009
Snake 1977,1989,2001,2013
Rooster 1981,1993,2005,2017

Dog 1970,1982,1994,2006
Tiger 1974,1986,1998,2010
Horse 1978,1990,2002,2014

Pig 1971,1983,1995,2007
Rabbit 1975,1987,1999,2011
Sheep 1979, 1991, 2003, 2015

Common Systems of Astrology

Each version of astrology deals with identifying cycles within given time periods and observing correlations to personality traits in individuals born within specific time segments of those time periods called signs. Perhaps the ancient astrologers observed the celestial connections as a two-way transfer? Could it be, that as the celestial bodies and celestial time periods impart an effect on our personalities the ancient storytellers in turn projected human qualities even mythical identities back onto the constellations? Perhaps lending further meaning to the old saying "as above so below".

Another powerful common claim of all of these systems is that the cycles have meaning and that certain time periods within each cycle correspond to human traits. Many traditions have and still practice a deliberate timing of events by correlating them to specific planetary alignments and time periods. According to legend Alexander the Great's mother was told to try and hold the birth moment to a precise time of day. And even in present day China many future parents wish to have children born in the year of the dragon. Is it all just superstition? Or could it be that these practices are based on something that can be demonstrated using today's science?

Another common thread is that all of these systems are calendar based, calendars that are based on either the Moon or the Sun, but not the stars. From the ancient Vedic (Indian / Hindu) astrology , to the lunar the based Chinese and Mayan systems and the solar based systems of ancient Kemet (Egypt) and Sumer now practiced in the west, they all have the common thread of the celestial bodies within our local solar system at their core, not the distant stars.

It would require an entire second book to go into the details of how the different astrological systems are related. For now it's enough to emphasize that it comes down to the planets and the sun interacting, resulting in a planetary system with the Earth as the center and the focal point.

Common Systems of Astrology

Clearly, these astrological systems are Earth sciences based on the interactions happening in our solar system with the Earth as the center. It is a science that we are more or less in the dark as to how it works when it comes to us as individuals. The existing astrology systems are only remnants, partial explanations of what may have been fully developed sciences at one time. There is a huge amount of evidence that shows advanced knowledge existed in ancient times, from stone building technology to astronomy that defies current explanation. It is very possible that ancient astrology is one of those previously advanced practices that has mostly been lost and that what currently remains is fragmented and mired in superstition.

If the claims of astrology are true, then the true origin of personality traits, the different forms of intelligence, and the source of talents come from the sun and the planets. So next we begin to venture deeper into the connections that are indeed present between ourselves and the planetary forces around us, and introduce concepts that begin to explain and uncover how these influences affect us all. In the Part Two we will be introduced to a theoretical model, backed up by scientific evidence, that may push our understanding of our connection to the planets even further than we previously thought possible.

Recommended Reading:
Secrets of Chinese Astrology: A Handbook for Self-Discovery, by Kwan Lau
The Handbook of Chinese Horoscopes, revised edition, by Theodora Lau

PART TWO
SEEING THE CONNECTIONS

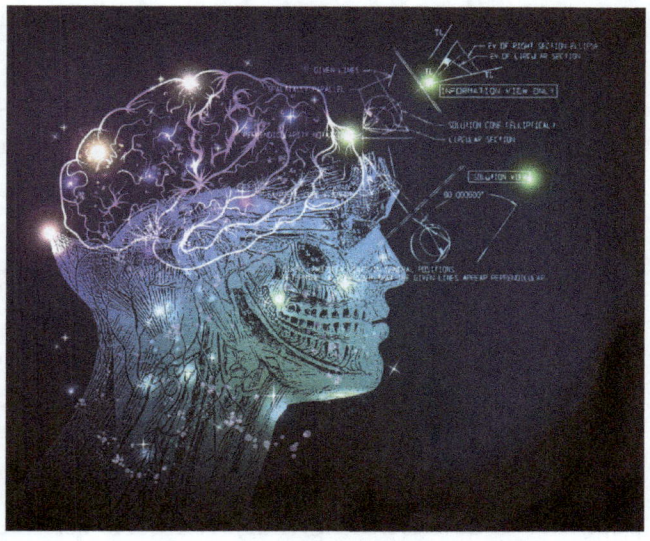

We have been presented with a great mystery. How do the planets affect humans? Astrology claims that the positions of the Sun, Moon and the planets, influence and even shape our personalities at birth. That's an incredible claim that requires incredible proof. In order to understand how the planets affect us, we first have to understand how planets affect each other. Next, this effect would have to affect us. To help us get closer to answering this we need to explore two major mysteries - first the mystery of how the brain stores information and second the mysterious subject of gravity.

Is there some "special material" in our brains that interacts with planets? If so how do we store this effect at birth to give us our individual personality? The brain is obviously the control center of behavior, so we will start our search there. To enable us to see new connections we have to look deeply into little know areas of neuroscience and explore the riddles of gravity, and see these fields as connected, not unrelated and distant topics. Are we mysteriously connected to heavenly bodies? Does the motion of the solar system really affect our personalities? Let's find out if it's true.

SECTION 1
CONNECTING THE PIECES

Now we begin our search to find the evidence that our personalities come from the Sun, Moon and planets. And let us not forget that "we come from" those giant spheres moving out there in space. Imagine yourself witnessing time reversing, as all the things around you turn back into giant clouds of dust returning to the center of a star, a sun like ours, just before it explodes. Then the idea of our personalities coming from a cosmic source doesn't seem so impossible anymore. What we have now is a few key parts of a whole picture that we need in order to begin to understand the entire life system that is going on around us.

With the Earth's axis as the centerpiece, and the planets aligning with the Earth, everything now points to gravity being the link of the solar system. What we have seen in part one was a description of a planetary science with the Earth as the central piece and with each of us being affected by the planetary cycles. What do we already know that could make this possible? The first thing that even basic science willingly admits is that we come from stars, our bodies are literally "star dust". And biology tells us that our bodies interact with sunlight, all life needs sunlight. And of course, there is the Moon and the tides and maybe there's a connection to water thrown in. Before we begin to sift through clues, let's gather a list of expertise that we will need to solve this puzzle.

To start, we are going to need information about the solar system and how the planets interact. That leads us to gravity, which modern science cannot fully explain, so the answers we will get from mainstream science are going to be limited. Knowing that the Earth's axis and planetary motion in the solar system is central to astrology we have to take a closer look at gravity, the only known force capable of affecting the entire system. It's primarily the astronomers and physicists that claim astrology is false, but they don't study the human body. So there is a basic disconnect here.

Who can put the complete picture together and link the human body to the planets? For the link between us and the planets, we are going to have to borrow from multiple sciences and bring them together to get answers. But in the end, we will rely heavily on neuroscience because they know the most about the human brain, and if the planets influence our personality the brain is a good place to start.

Connecting the Sciences

What we are quietly stating is that the brain is somehow capable of interacting with gravity. Could unrelated fields hold the keys to each other's long standing mysteries? How many physicists are looking for the missing link between gravity and magnetism, and electricity in the field of neuroscience? And, how many neuroscientists are looking for clues to the brain's mystery of memory in the field of astrophysics? Why would they? They have their long held methods that have worked for so long.

Kenemonics is a new of looking at old questions. It puts together the facts connecting astrology, and its birth chart, with personality traits, astrophysics and neuroscience. Also, we must include genetics as a key part in getting a complete picture of the entire human puzzle.

"Every living being is an engine geared to the wheelwork of the universe. Though seemingly affected only by its immediate surrounding, the sphere of external influence extends to infinite distance."

Nikola Tesla

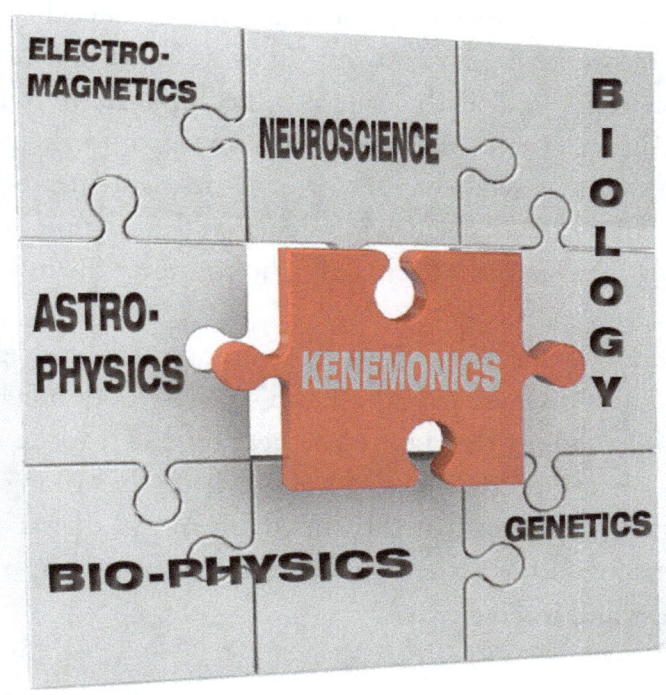

Over the course of the next section we will begin to explain each part of the model starting with the System of Planets and how the planets interact with the Earth as the focal point. If astrology's claims were obvious to see given standard biology and physics, then someone would have explained it already. The explanation not only needs multiple areas of science, but will push the boundaries of all of them. Just like genetics and neuroscience have created new frontiers beyond biology, both have relied heavily on computer science and modeling to make breakthroughs.

Kenemonics

The Kenemonic Model puts all of the pieces together in a very simple manner. Astrology has provided us with clues to the beginning pieces and the behavioral evidence for the last piece. But the pieces in between remain mostly unknown. The birth chart is our first big clue, it gives us a great starting point. It tells us that the position of the planets at birth with Earth as the focal point, cyclical changing configuration, affecting the earth like weather. This model is theoretical but based on evidence that begins with the system of planets that make up our solar system. The planets orbit the sun in cycles due to the pervading force commonly known as gravity, these cyclical movements cause fluctuations in the gravity throughout the solar system, but model's concern is how this affects the Earth.

A few scientists question what can a planet do from such a long distance? Carl Sagan went on the record saying far away gravity was simply too weak to affect a human at birth. The problem with that logic is that the Sun and the Moon have tremendous effect on the Earth. Even at such great distances we see the tidal effects, so we know their gravitational forces are getting here. So the next question is what happens when these gravities overlap as the earth changes alignment with them? This is key to the new model. Overlapping fields are almost unheard of in classroom science, we study interference patterns of light but never the interference of fields like gravity or magnetism.

Do those fluctuations show up on earth as changes in Earth's gravity field? It seems likely. We will also discover that the Earth's gravity does much more than just pull things to the ground. These fluctuations that take place here on Earth go on to further interact with the human nervous system, most specifically the brain where the arrangement of planets at the time of birth is permanently stored. It is this stored "memory" that is contained deep in the structure of the brain that astrology's birth chart describes as an expression of individual personality, a specific range of behaviors and interactions.

Still, astrology's basic idea that celestial bodies imprint and influence an individual's personality is often ridiculed as absurd. If you think of human beings as existing and being composed of every basic force we know, electricity, magnetism, and many interactions with light. It's not much of a stretch to include mysterious interaction with gravity among the list of what we know. In fact, why would we leave gravity out at all? If you look closely at astrology's claim of planetary aligning, this is an obvious reference to gravity.

Connecting the Pieces

1: | BABY BORN – INSIDE THE SOLAR SYSTEM |

TARGET EARTH:

ALIGNMENT OF PLANETS

ALL HUMAN BEINGS AFFECTED

2: | STORED IN US |

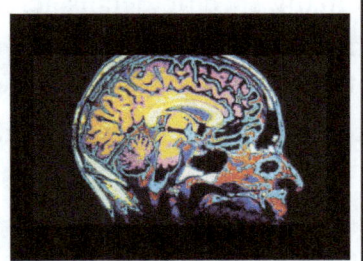

BRAIN FROM BIRTH & BEYOND:

AT EXACT TIME & PLACE OF BIRTH WE ARE IMPRINTED THROUGH THE UNIQUE QUALITIES OF THE BRAIN AFFECTED BY THE SUN, MOON AND PLANETS.

3: | YOUR LIFE |

PERSONALITY TRAITS FOR LIFE

ALSO INFLUENCED BY SHIFTING PLANETS POSITIONS

Where Does Personality Come From?

The question of personality is a three-way debate. One group say genes, the second says astrology knows best, and a third group that says it the environment you are raised in. Of course, there are other views on personality out there, but they don't have the evidence of the top three. Personality is a set of behavior traits of an individual person. Next we look at each of the three arguments and follow the evidence.

Our DNA, or genes, are the blueprint to build our bodies while the solar system gives us our basic personality traits at birth, and the environment and our experiences develop and influence the body and personality. All of the those combine to make each of us totally unique individuals. But genetics with two DNA samples cannot tell us how to match personalities between two people. Genetics cannot tell us about personality, give a sample of DNA. Without a doubt, genetics has shown extraordinary prowess in matching physical traits given related samples, like paternity testing and forensics. But genetics has shown nothing outside of physical traits.

Astrology's birth chart gives many specific details across each area of life given only the exact birth time and place. By showing individual personality traits astrology has been able to create a basis for analyzing compatibility. Genetics and behavioral scientists may not even attempt to deal with compatibility. Both fields cannot tell you anything about individuals without interacting with them first.

A part of the third group tries to push the idea that 100% of personality is random and unexplainable. Personality traits are as real as physical traits. Rare traits like vision, analytical and creative genius, leadership ability. Even less rare traits like, introverted, extroverted, and even-tempered astrology claims to know the source. We simply need the birth information, then it's all in the birth chart. We can debate the necessary environment needed to cultivate traits, but they are real in life. To those who argue that there is no built-in personality, there are two simple questions, how do you explain the variability we see in human personality types and traits, and what is the basis for compatibility?

The bottom line is that we all know that we have distinct personalities, we are each a unique mix of experiences, memories and traits that we have shown since the earliest days of childhood. This fact cannot be explained away as just some random accumulation of experiences. It's obvious that personality traits are real. To dismiss them is silly, almost stupid. So it is really not a matter of whether or not we have personalities, we do. The question remains, where do our personalities come from?

POSSIBLE PERSONALITY SOURCES			PERSONALITY EXPRESSIONS	
Kenemonic	GENETIC Physical blueprint	Environment & Experience	Individual Traits	Basis for Compatibility
CASE 1 ✓	✗	+	✓ Explanation available	✓ Explanation available
CASE 2 ✗	✓	?	? No Explanation	? No Explanation
CASE 3 ✗	✗	✓	? No Explanation	? No Explanation

Case 1: Kenemonic
There is a "built-in" personality, as described by Astrology, it is "somehow stored" in the brain at birth and both Experience and Environment also play vital roles in its expression.

Case 2: Genetic
The genes somehow code for personality, although no behavioral genes have ever been identified in the same definitive manner as genes that code for physical traits have.

Case 3: Humans have no Personality
Intrinsic Personality is a simple illusion, only the immediate environment and experience set are factors in human behavior. Individual variability and compatibility are therefore random or unexplainable.

Using Models

There is a long tradition of creating and developing models to aid in discovery. I believe that by having a full model rather than just a theory, the process of gathering evidence through observation, and testing ideas is easier, than just a stated theory alone. This Kenemonic model is highly visual, that helps explain the parts that we can all observe for ourselves. A simple model also means anyone can add to the sharing of fresh ideas and viewpoints. Physical evidence is often unpredictably found, but ideas can still be developed and any time in the process. A model creator must be daring, to make that leap into the unknown and try to explain a phenomenon, often with very little to go on. All we can do now is gather all of the evidence and counter evidence, continue to observe and bring together a more complete picture over time. Anyone with the curiosity and will can join in. It's exciting to be close to a huge discovery!

   A prime example of the use of models in the discovery process is the atomic model, which started out with many "way-off" models, with some scientists proposing models that the atom was just a random "soup" of tiny particles with no structure. But through a long sequence of theories, and testing those theories, the actual atomic structure became more and more evident, revealing the stable nucleus comprised of neutrons and protons that was orbited by electrons in distinct orbits called orbitals. The discovery of the structure of the atom led to today's electronics. Atomic structure allows us to understand what makes certain elements conductors and others semi-conductors which eventually gave us the microchip. The usage of models highlights the importance of creative and imaginative ideas followed up by a continuous process of testing those ideas.

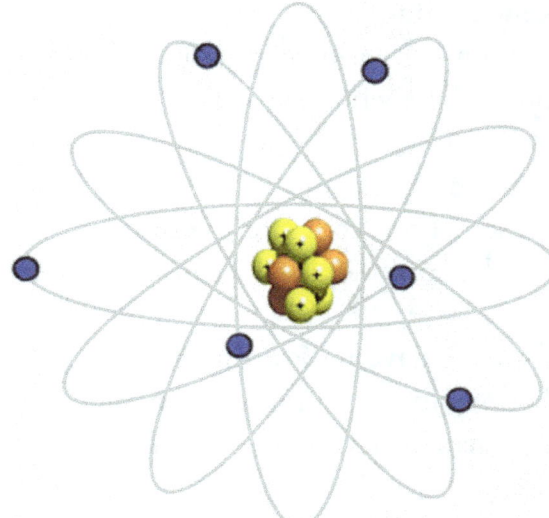

SECTION 2
The System Of Planets

   More of astrology's secrets have been unlocked. We now know that astrology is basically a calendar of the earth's tilt as the earth aligns with the sun and planets in the solar system. What is the hidden link between the Earth and the human; from planet-to-planet link to planet-to-people link. Gravity is the central force in the system of the planets.

   Gravity has always been a mystery to modern science. With the exception of the recent discovery of gravity waves, today's scientists have thrown in the towel on gravity and skipped past it to the new theory of everything, the theoretical intangible particle know as super strings.

   Many simple observations have been passed over by mainstream science. For example, the idea of gravity being a pull is incomplete. Gravity is much more than bodies in space pulling on each other or bending space time. There is a lot missing, many other observations that are key parts of a more complete version. The movement of planets is like a symphony of the solar system – not noise, but organized. Together, the planets create distinct patterns that blend and produce unique harmonies.

The Mysterious Force of Gravity

Gravity is sold to us as a force that is a pull, but that doesn't actually play out when you take a look at the facts. What is currently well understood about gravity is our ability to measure and calculate its force and the planetary motion due to gravity. Once you start asking basic questions about gravity it turns out we have more questions remaining than answers.

If gravity was a simple force that was only a pull then it would do just that and all the planets and the sun would pull each other into one giant ball. But that doesn't happen, why? Because the pull is just a surface component of gravity and we have the famous story of Isaac Newton's apple to thank for perpetuating this incomplete idea.

Gravity has more to do with form, shape and balance of motion than it does with pull. So much is left unexplained, so Modern Science simply does not deal with the questions in the university text books. There are apparent problems between what they tell us it is and what we actually can observe. The same thing they do with magnetism.

A huge missing piece in the gravity puzzle is, why do the planets remain separated? It appears that gravity has a balancing component that hasn't been given the same attention as its pull component. Also nothing seems to be able to block gravity. The largest influences in astrology are the sun followed by the moon, an obvious fit for the top two gravitational forces on the earth.

Gravity Miseducation

Gravity is more of a link or a tethering between "everything" we know. And there is a profound order to it. Is it that mainstream science isn't aware that gravity is much more than a pull or is this information simply not being shared? Objects are "pulled" toward their most nearest massive object.

Mainstream science wants us to think gravity is only a force. Cleary magnetism is a force, pulling and pushing magnetic objects, but there is also a field effect, and a little known sensory effect on many animals (which we'll cover later). For now let's keep the analogy simple, we'll use air. Air can be a force, it can move sail boats, make trees sway, or you can blow the dust off a countertop. But on a more subtle level slight vibrations within air give us sound, a transmission of information.

Let's move on to other forces. Electric and magnetic fields together they can transmit vibrations from a cell tower to your cell phone or radio tower to your car, but you'll never hear anything without an antenna. Does gravity have fluctuations? It has to, the alignments of planets are always changing. If gravity is more than a force simple force what is the antenna? We must have a way of picking up gravity's fluctuations.

Gravity fits the description of what the planets are doing in the birth chart. Gravity goes through everything, even other planets. Gravity also varies with different planetary alignments, such as a full moon and a new moon.

Let's compare gravity with its cousin magnetism. Magnets have a selective effect, like a magnet under a glass table, it won't affect wood or plastic but it will move another magnet. And there's a field to field effect, like fields repels, unalike attract. But nothing seems to act differently with Gravity! Or does it? Maybe we just need the right combination of fields.

Basic Gravity Rules we can see:
1) Every planet and moon has a primary tether.
2) Scale to Scale, the true power of gravity. Its property of acting over massive distances and on massive Objects.
3) Gravity Disc is caused by spinning, also gives direction.
4) Link or Tether to nearest massive object – Primary link
5) Massive to massive very noticeable, massive to small object very noticeable, small to small not noticeable.
6) Gravity is a system of order – orbiting connected to sometimes spinning
7) Not blocked by anything
8) Strength of pull varies with size of object and distance.

GRAVITY

Here we see examples of gravitational bodies producing orbital planes or "gravity discs". As suns and planets rotate they create these disc shapes at their center. The theory of bending space time does not fully explain this. The illustrations below show the natural force of gravity, the ordered arrangement of orbital planes like our solar system and rings around planets. Notice how "nothing" naturally orbits with the red dotted line, opposite to the orbital plane. This happens on the planetary level, the solar system level and even up to the galaxy level. In all of these cases there is a huge gravitational center. Could it be that spin adds a component that allows for balance and separation? There wouldn't be much of a solar system if gravity were only a pull, there has to be other components; the spin and resulting disc and separation are evidence of this.

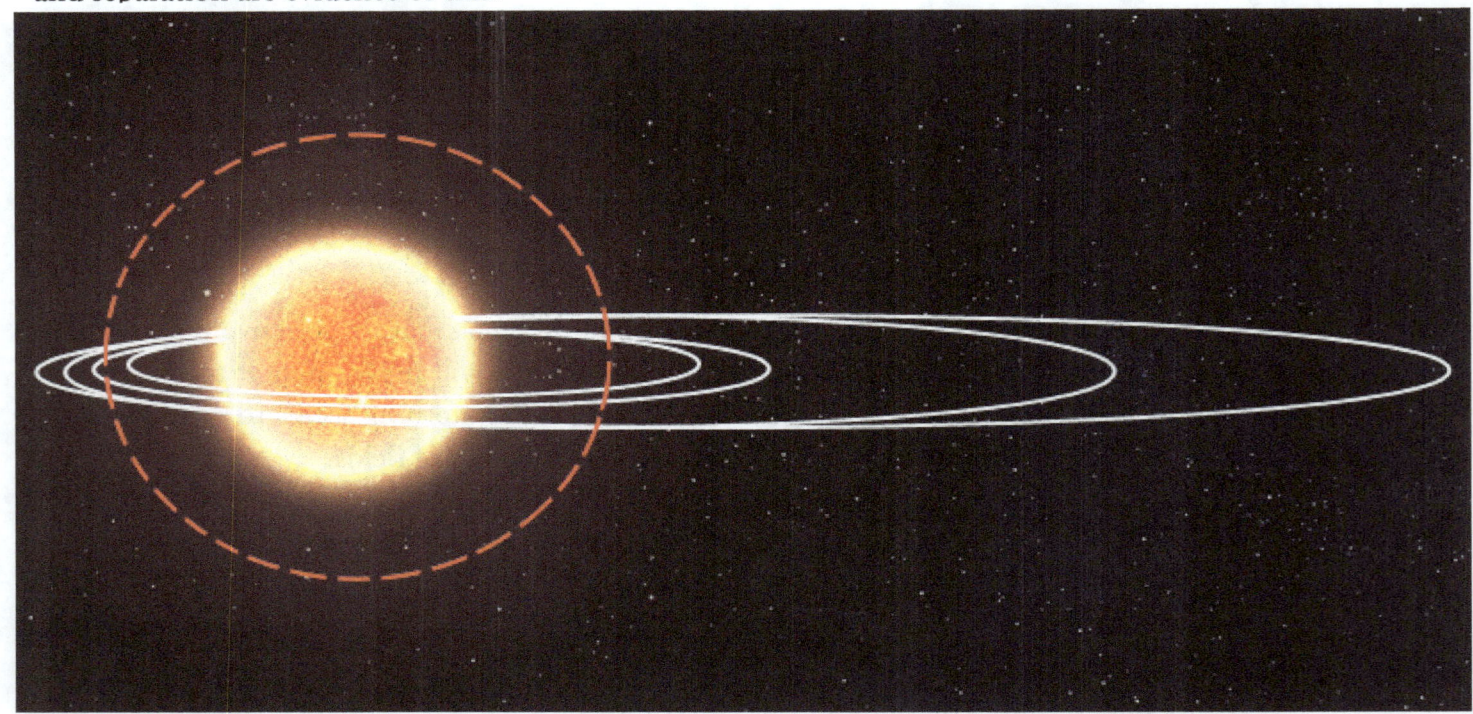

Figure (above) The solar system is disc shaped.   Figure (below)

GRAVITY DISC

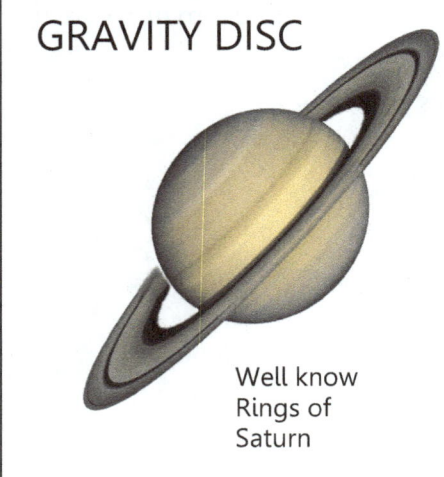

Well know Rings of Saturn

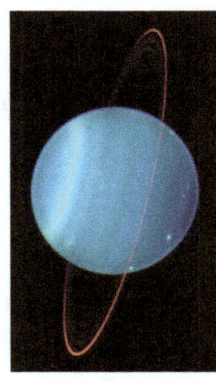

The rings of Uranus: Uranus spins on its side relative to the other planets. Neptune, also has rings.

Galaxies also spin, in a spiral shaped disc, with a massive black hole at their centers.

Within our solar system we can readily observe two basic types of gravitational bodies, that being spinning and non-spinning gravitational bodies. This difference may be important, even outside of the solar system, because of the fact that spinning celestial bodies seem to produce these orbital planes this is not limited to the rings of planets or the orbital plane around our sun. In fact, this is seen on the level of solar systems in general and also entire galaxies that rotate in this disc shape.

Conversely, less stable bodies such as asteroids and comets may rotate, but they do not spin on a stable axis. But why is the distinction of spinning important? This distinction is important for the apparent balance that seems to exist between and around these spinning bodies. The vast majority of the observed celestial collisions involve these non-spinning bodies. When a non-spinning body interacts with spinning body the non-spinning body, such as an asteroid, is often subject to the spinning body which can result in orbit alteration and collision. If there is indeed some "special" interaction occurring between spinning bodies this would explain the separation that exists for the most part between these bodies when they exist on the same scale such as galaxy to galaxy and planet to planet.

The suggestion is that spinning bodies have "special interaction" with one another and possibly reveal another component to gravity beyond the "simple pull" that we are commonly told. How do these planetary forces interact? We can no longer take for granted that gravity is a simple pull. This raises a further question.

What is the nature of gravitational fields interacting? That is to say what happens when these planets gravitational fields in our solar system line up or align in certain ways? What are the top aspects of Gravity. Hemisphere in Spinning bodies – produce discs at the junction – equator Pull The blocking and interacting problem, nothing blocks it, but does it resolve in any way? No other forces act in this manner. So one of the most important things to study would be what happens when gravity a interacts with gravity b? What happen at the point of intersection?

Patterns of Interference & Overlapping Fields

With the exception of optics, the topic of overlapping energy fields and waves is not covered in modern physics text books. In order to explore gravity the concept of overlapping fields is a potentially the biggest key. For one, the fact that gravity between many orbiting bodies appear to establish a form of equilibrium. Then there is the spin component of gravity, this may help provide stable orbits of moons around planets and planets around suns.

We need to consider what happens when the gravity of two planets overlap. Overlapping must produce patterns. So then, what results from the patterns of overlapping gravity fields? We all know two objects cannot occupy the same space, but clearly fields can occupy the same space! Astrology's birth chart leaves us some clues. The birth chart shows planets aligning and producing effects we see in behavior. It now seems clear that the aligning of planets is the overlapping of planet's gravity. This overlapping creates distinct patterns, shown in colors below. What is happening with planets may be similar to light holograms, were overlapping light creates image patterns that are able to be stored. Amazingly, we may be able to "pick up" the patterns produced by planets aligning.

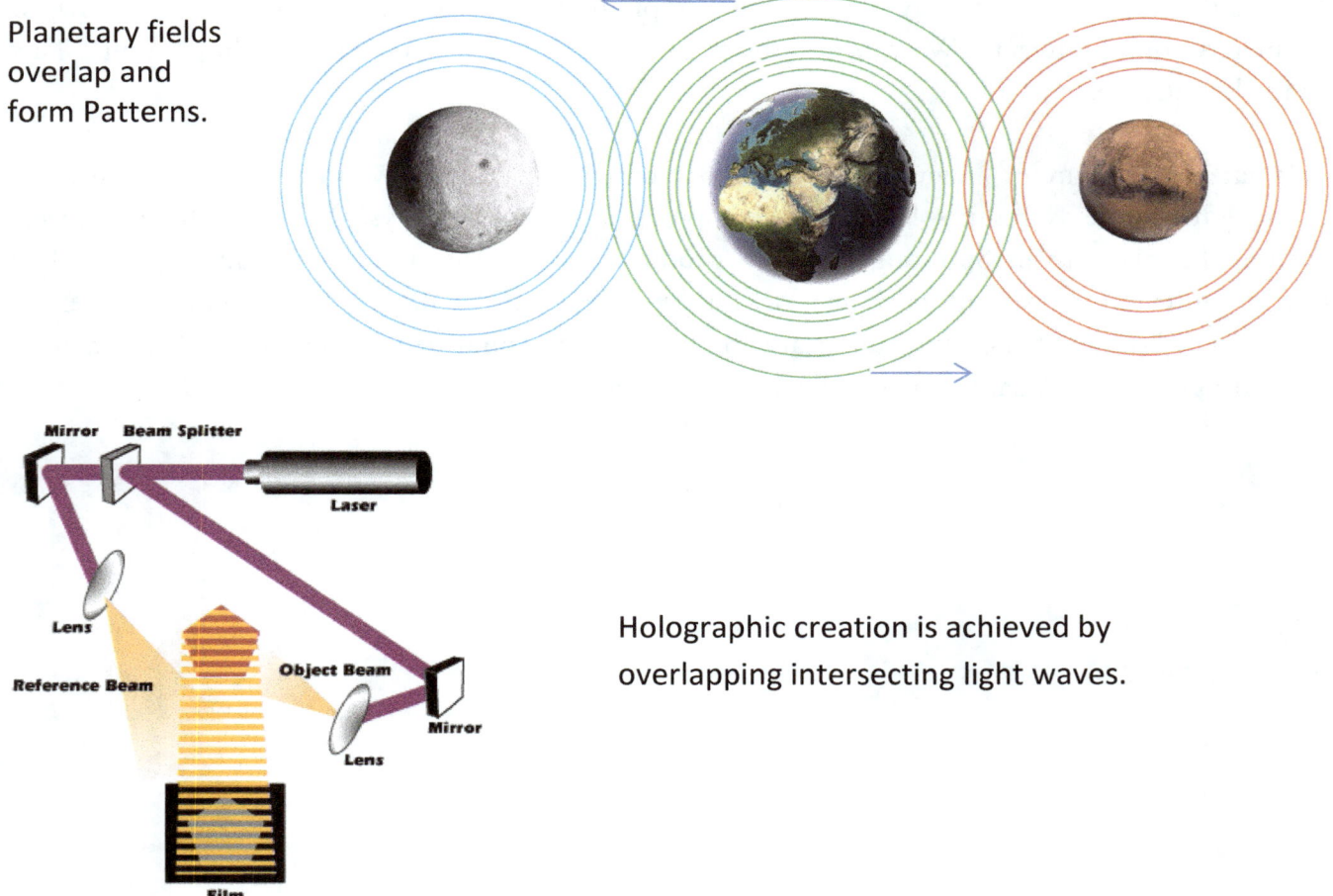

Planetary fields overlap and form Patterns.

Holographic creation is achieved by overlapping intersecting light waves.

PLANETS IN CONFIGURATION

Planetary alignments are vital to showing additional components of gravity, we normally don't consider. There are many planetary factors; size, relative size, distance from Earth, and spin, to name the main ones. Given these factors the Sun then the moon are the top two forces that affect us. This is totally consistent with astrology's birth chart. Each one of us is at an intersection in space for all of these converging planetary alignments.

So the next question is, how are we affected? The human brain and nervous system is a good place to start. A neurological effect that happens on a deep sensory level, would make astrology's claims of influence on personality traits valid. But there is also a geological effect that goes well beyond the known tidal forces, it is this effect and the relevance of planetary alignments that we will explore first. Just like still air or water, when disturbed you see or hear the effect. The same with gravity, these alignments cause large changes.

Examples of typical "Alignments" generally meant to describe when 3 or more celestial bodies form a line (or an angle).

PLANETS IN CONFIGURATION

Astounding research of solar system alignments shows a powerful correlation between large seismic events and alignments. The data is so compelling that the research for this book expanded. The data does not only give the exact day but it shows an approximate 2 to 3 hour window following an alignment then the seismic event occurs. All of this is a simple matter of showing the data, there is no complex theory behind it, the data speaks for itself. The alignments causing the quakes are linear, 180 degrees, just like a full moon. Astrology calls these oppositions, said to be very strong among the various types of alignments. The implications of this research are tremendous. Perhaps this could be used as an early warning strategy, a possible warning broadcast of times for the likelihood of major earthquakes?

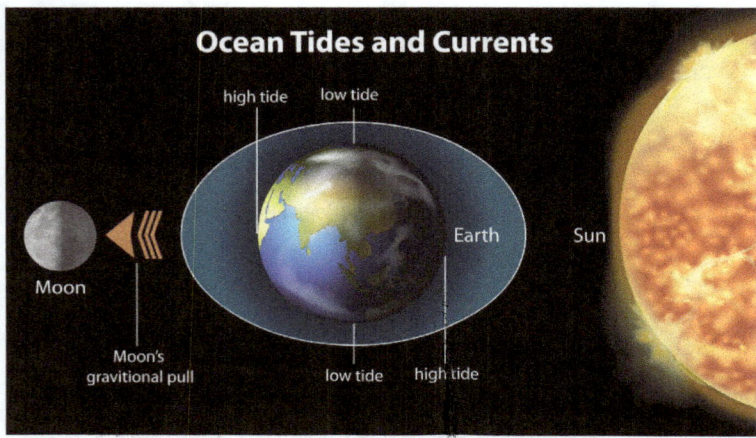

It is known and accepted by mainstream science that ocean tides are caused by the Moon and Sun pulling on the ocean water. Have to disappoint a commonly help belief that the moon effects water, it DOES NOT affect water in any special way. The MOON Effects oceans, which are massive. Massive objects pull on each other in noticeable ways. But the moon will not affect water in a glass, your bathtub or in you.

Like water in a bowl, our ocean water can move independent as one. The same goes for our land on earth, there are huge gigantic plates of land that sit on top of molten rock they also can move causing earthquakes and tsunamis.

So if the sun and moon can pull the oceans, then why not pull the giant sections of earth? They do, and there is plenty of research that shows that this does happen.

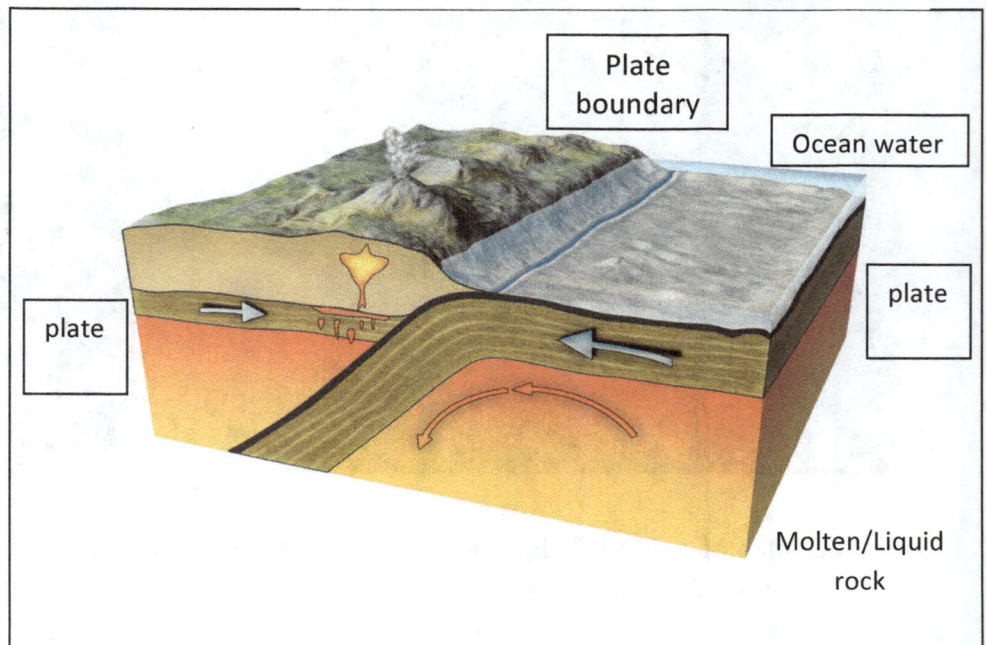

*It is claimed that physical tidal forces due to gravity, such as during a full moon, may have an effect on humans but that physical force is extremely small, if it is present at all.

The System Of Planets

How can such a massive effect take place between celestial bodies and transfer to the Earth's crust plates yet we as individuals are not affected in any way? Or to put it simply, why don't we feel this transfer of force when planets align? The reason has to do with scale and frame of reference. Massive celestial bodies are for the most part on the same size scale, oceans and crustal plates are themselves gigantic masses, at this scale interactions of huge forces are easy to see. But human beings are only secondarily affected being on a smaller frame of reference due to our physical size. So the plates to move first, then we are in turn moved by the seismic activity. It's indirect, not direct because we are not big enough to interact directly with celestial bodies when it comes to physical force.

But bear in mind, gravity has several components, physical force transfer is just one. The component affecting us is most likely at the field level. We will cover a simple example of this shortly. But first, a quick explanation on why size matters when it comes to interactions with physical forces. So it seems, gravity has a force component that acts at a distance directly affecting masses on relatively the same scale of mass, while also having a field component that selectively interacts with our nervous system.

Japan
March 11 2011 **9.0 Magnitude** quake was predictable using **alignment** data.
See APPENDIX A

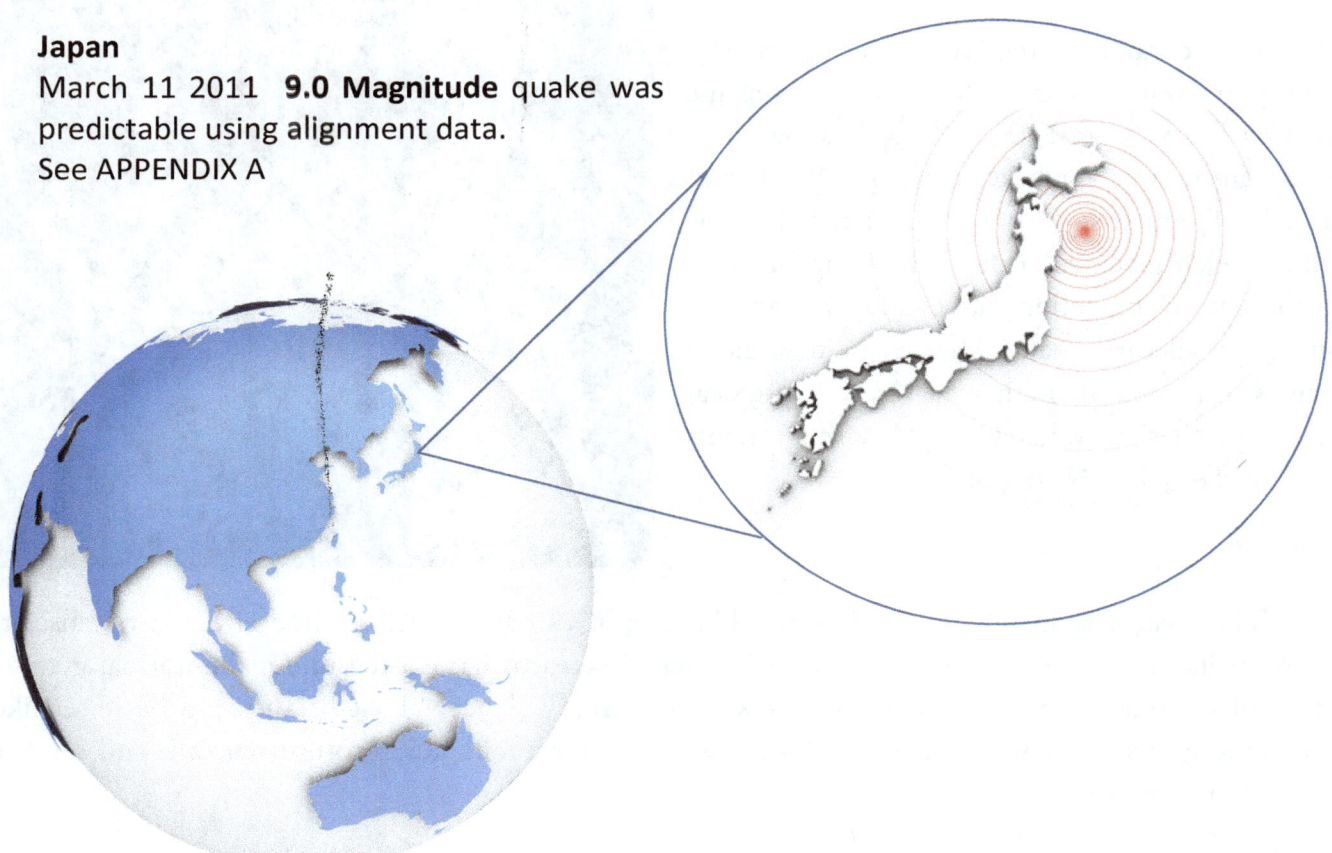

Planets In Configuration

Extraordinary events have been happening in our solar system with major effects occurring on Earth "right in front of us". One of the first major alignments occurring between the comet Elenin, the Earth and the Sun on February 27th, 2010 caught the attention of few keen observers who drew correlations between this alignment and the major earthquake in Chile on that day. But it wasn't until the event on March 11, 2011, which was forecasted, that more widespread attention was generated. Even more incredible, is the data compiled data that show that certain types of planetary alignments not involving any "runaway" comets may have serious effects on the Earth.

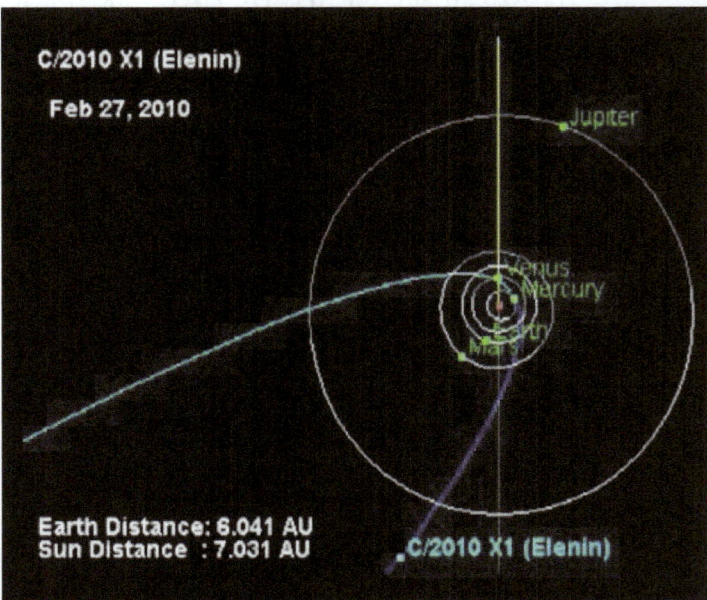

The data is extensive, covering mostly recent years but also extending back to seismic events as far back as 1906. Records show clear correlations between significant seismic events and celestial alignments. It is important to be cautious, even given such data. Keeping in mind that there is no suggestion that celestial alignments alone cause all earthquakes, rather the data alone provides evidence that there is a real physical connection to the events when these alignments do occur. See APPENDIX A.

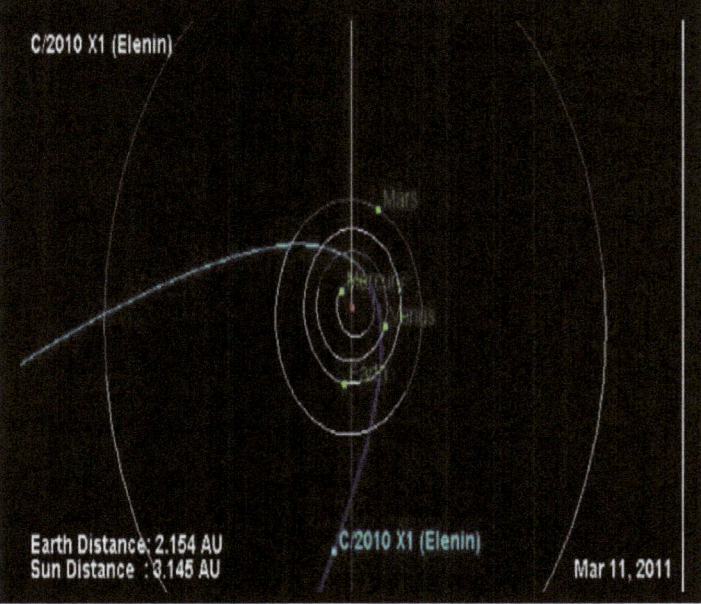

The basic reason why a small comet like Elenin can have such a large effect is because it is somewhat like the straw that broke the camel's back. The earth has existed under the normal gyrations of the solar system, sun, moon and planets with a normal balance. That balance can be upset like a perfectly weighted see saw, balanced on both sides. It doesn't take much to unbalance it and send one side crashing down.

CLOSER TO THE ANSWER

The picture below is the magnetic field of the Earth blocking radiation from the Sun. The boundary is called the bow shock. Because of the Earth's Iron Core, our planet has an enormous magnetic field covering and going through the planet. The Earth's magnetic field is large in size but relatively weak in strength. The magnetic field is used for navigation giving us true north and south direction on a compass.

Previous researchers have thought the earth's magnetic field was behind what astrology is describing. The Earth's magnetic field is global. But magnetic fields just don't have the ability to reach us anywhere on the planet at birth or after birth. So by process of elimination, we can eliminate the earth's magnetic field as being the force that affects us like astrology describes. Still, we need to confirm we are on the right track with gravity. And since gravity can reach us anywhere on the planet, gravity fits what we are looking for.

The Earth's magnetic field can be:

1) blocked by special materials

2) overpowered in small areas by other magnets.

So for at least these 2 reasons the Earth's Magnetic Field is **NOT** the cause of any "astrological effects".

Gravity **DOES** fit what we are looking for.

1) Gravity cannot be blocked! Gravity goes through anything, even planets.

2) Gravity increases when planets align.

So now we are left to focus on gravity and how it can affect us?

THE ZODIAC REVEALED

We have seen how the Earth's axis aligns with other planets gives the zodiac. Now we can focus more closely on how this is projected onto the Earth's surface thereby effecting us. Knowing what we know about the natural occurrence of planetary rings and the gravity disc, it is easy to see the band of the zodiac is in the exact position of these naturally occurring discs at the equator.

Gravity spreads across the surface of earth differently, spinning in opposite directions on either side of the equator. This must affect all of us on the surface differently. The real zodiac corresponds to this natural ring or disc created by planetary spin and gravity. Gravity is hemispheric, and corresponds to the Earth's poles, axis and spin. The zodiac is not something made by a far away source, it is a local, and its center is the center of the earth. Because of the Earth's axis tilt the system remains constant because the seasons and the signs will always be there unless the tilt goes away. We visually picture how overlapping patterns of gravity would change as alignments change, season to season and sign to sign.

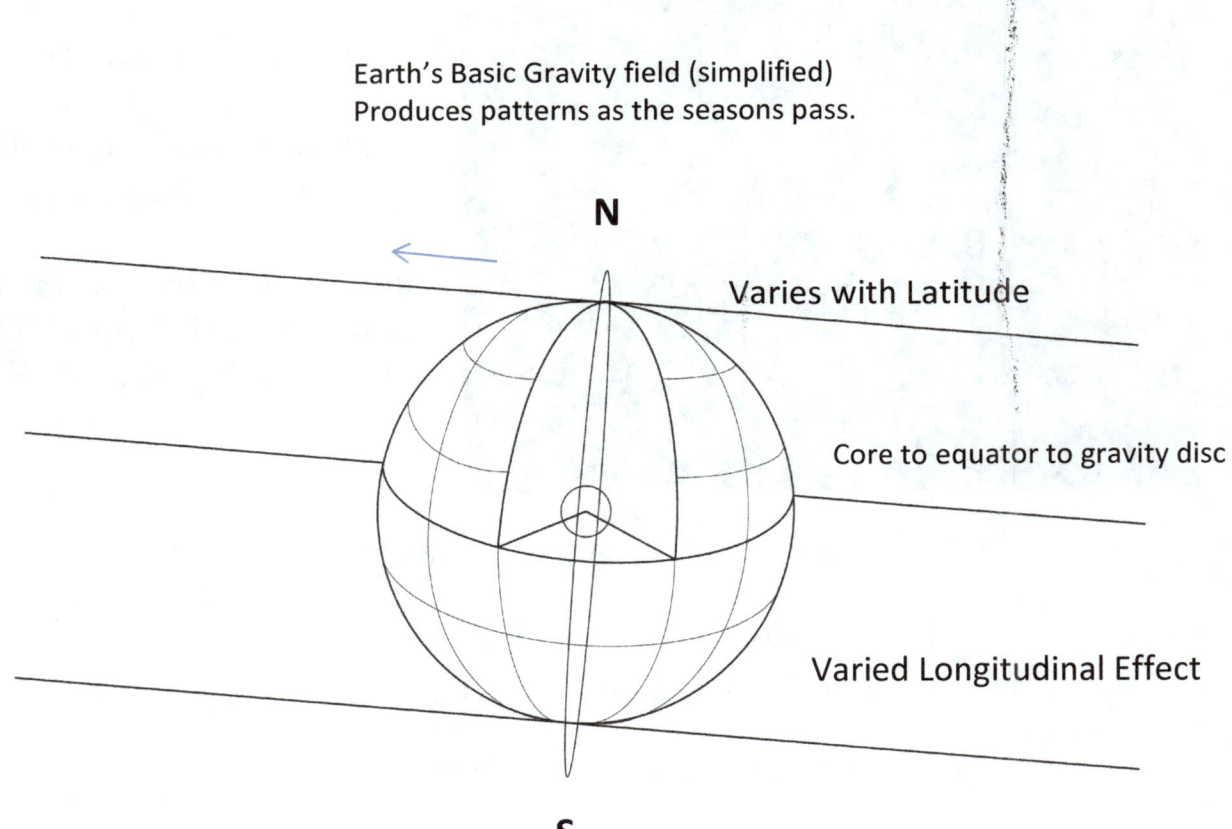

THE ZODIAC REVEALED

The surface of the Earth and everything on it rotates through this effect, we call zodiac, that extends from the Earth's core and pole. Alignments differ depending on a planet's position to the poles and axis tilt – we call these signs. A full revolution of the earth spinning on its axis observed as a day goes through every point of this field. In astrology, this is called the houses.

The birth chart really is a diagram of overlapping forces from planets aligned with the Earth's axis tilt. Like musical notes, these patterns have no problem overlapping and still produce a distinct effect. Human personality traits are predictable within astrology, here on Earth, but would change on other planets. If Elon Musk is successful with his Mars colony, the zodiac of Mars will shift and be quite different than what we experience on earth.

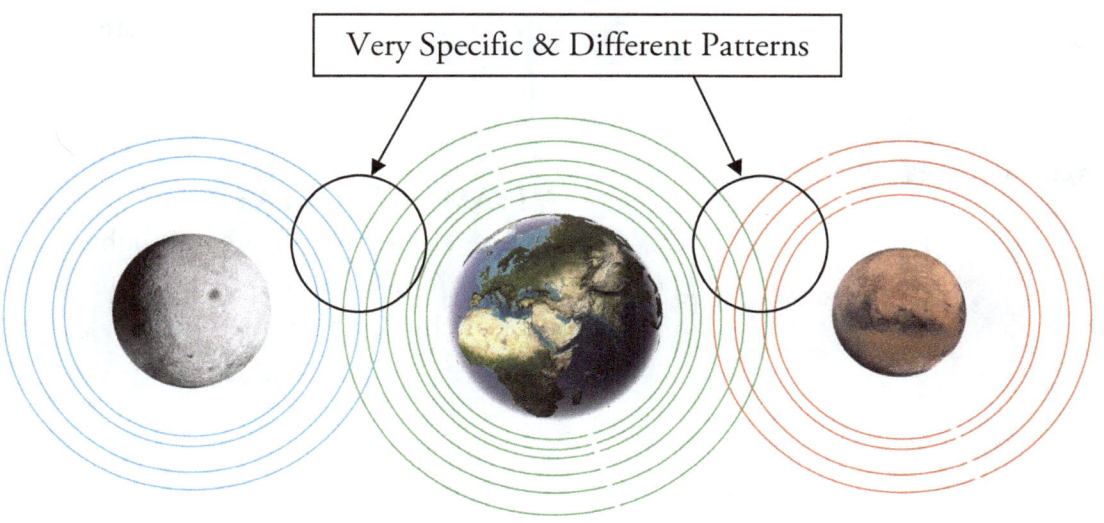

figure 65a   Mars & Moon patterns

THE MOON

Distance from Earth:
SPIN: – the moon does not spin
AXIS TILT: – the moon does not spin
MASS:
Relative Size: 2nd largest object in the sky. Second most powerful Object in Astrology.

MARS

Distance from Earth: 48.6M mi 78.3M km
SPIN:
AXIS Tilt: 25.2 degrees
Diameter: 4,219 m 6,790km
MASS: 6.42e23 kg

The Body Interacts

We exist here on Earth, adapted to the forces of nature. The force of gravity is no exception. Actually our bodily functions are in direct interaction with forces that come on a massive scale from our sun and earth. Our bodies being connected with the forces of nature, is a rule not an exception.

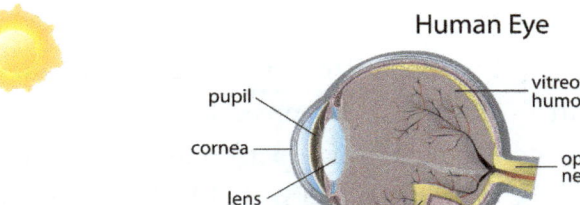

Eye Sight
We take this for granted, but we interact with light energy (electromagnetic waves) in the form of visible light. Nothing ground breaking, but still a fascinating bit of science is behind how our eyes and brain receive and process light into a view of the world and store that as memories.

Figure 66a

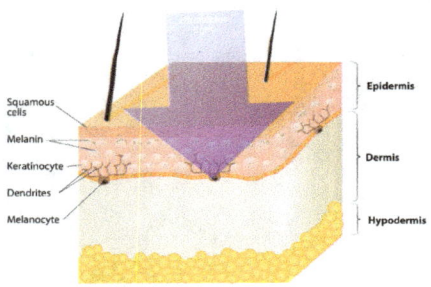

Figure 66b

Vitamin D creation
Our skin's ability to absorb sunlight and process it in Vitamin D happens without us being aware of it.

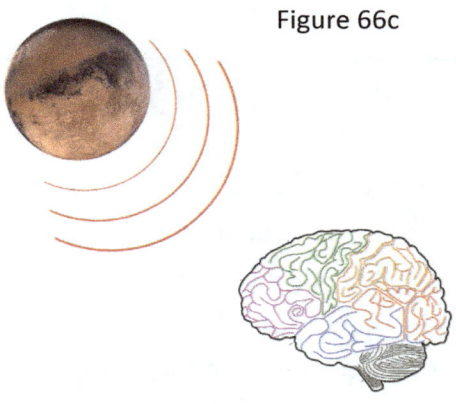

Figure 66c

Solar System Effect
Now let us focus on the brain. Just like vitamin D we are being affected without being directly aware. Just like our brain interprets light in to colors and images, our brains interpret the gravity of the sun, moon and planets into influences on human personality.

SECTION 3
THE ESSENTIAL BRAIN

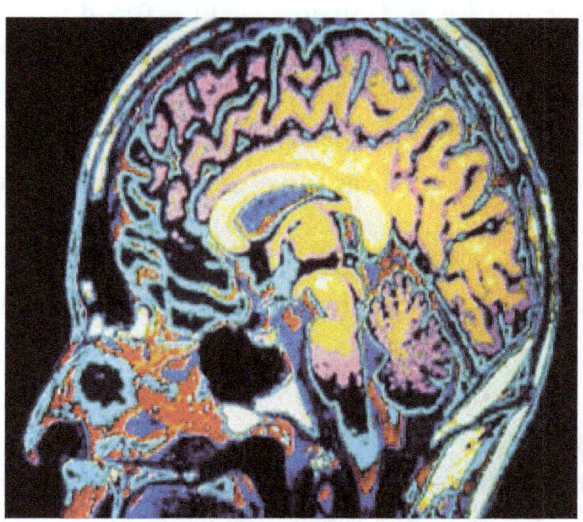

Do the Sun, Moon and planets in our solar system affect our bodies, if so what part? For our, search the most likely place would seem to be our brain. We focus on the brain because we know that the brain is the center of memory and behavior.

Our bodies have evolved to interact with our surroundings. It only makes sense that planetary forces would be a part of that interaction. Consider our ability to sense and interact with light, not just eye sight, but our ability to make vitamin D from sunlight. This is undoubtedly an example of directly connecting to the sun.

Now consider gravity as both a local earth phenomenon, and also a prominent force interacting throughout the entire solar system. Could it be that we have an unconscious ability to interact with gravity? To answer that, we need to begin looking for clues as to how gravity can affect the brain.

THE UNIQUENESS OF THE BRAIN

The brain is the most complex part of our body. Using the well-known five senses, the brain is the central sensor for the outside world. And if what astrology is saying is true, the brain is also able to sense the movement of the planets. As far-fetched as that may sound, the evidence and logic behind that idea is consistent. It is a matter of looking at the facts staring with the birth chart. The birth chart gives us details about personality traits given only birth time and place based on the position of planets at the time. This leads us to the brain, where personality is expressed. From there we can consider the possibility of the brain having another sense or ability, just an unconscious one. How better to explain that personality traits seems to match up with when people are born?

Inside the brain, may be the only known material to bridge what we know as gravity with the other forces such as electricity. There is no other device, organ, or thing that has so many electrical connections and electrochemical connections, in such a complex arrangement. The uniqueness of the sense organ for sight alone is extremely complex. It has to be admitted that mainstream science does not know everything there is to know about the deep inner workings and structure of the brain.

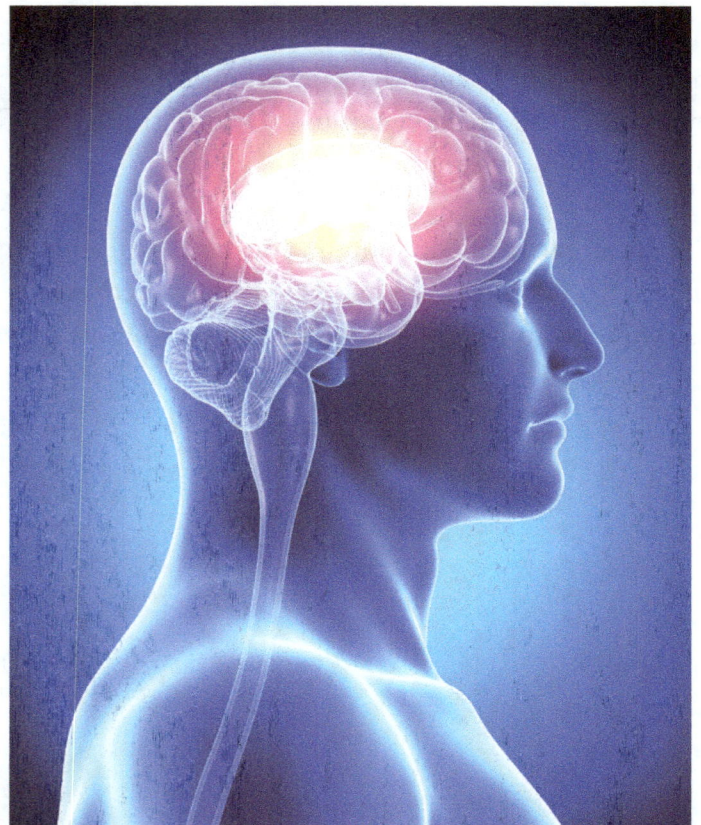

The Brain

Mainstream science has collected a massive amount of information about the brain but could it be that we still have only scratched the surface of what the brain is capable of?

The Essential Brain

Could it be, that one of the great mysteries that has eluded physics and great minds like Albert Einstein could be solved in the seemingly unrelated field of Neuroscience? It appears that within the brain the link between the fundamental forces of physics may be found. Furthermore, gravity is the only known interactive force that goes through everything including planets. And it is the only force that is compounded by the alignment of planets.

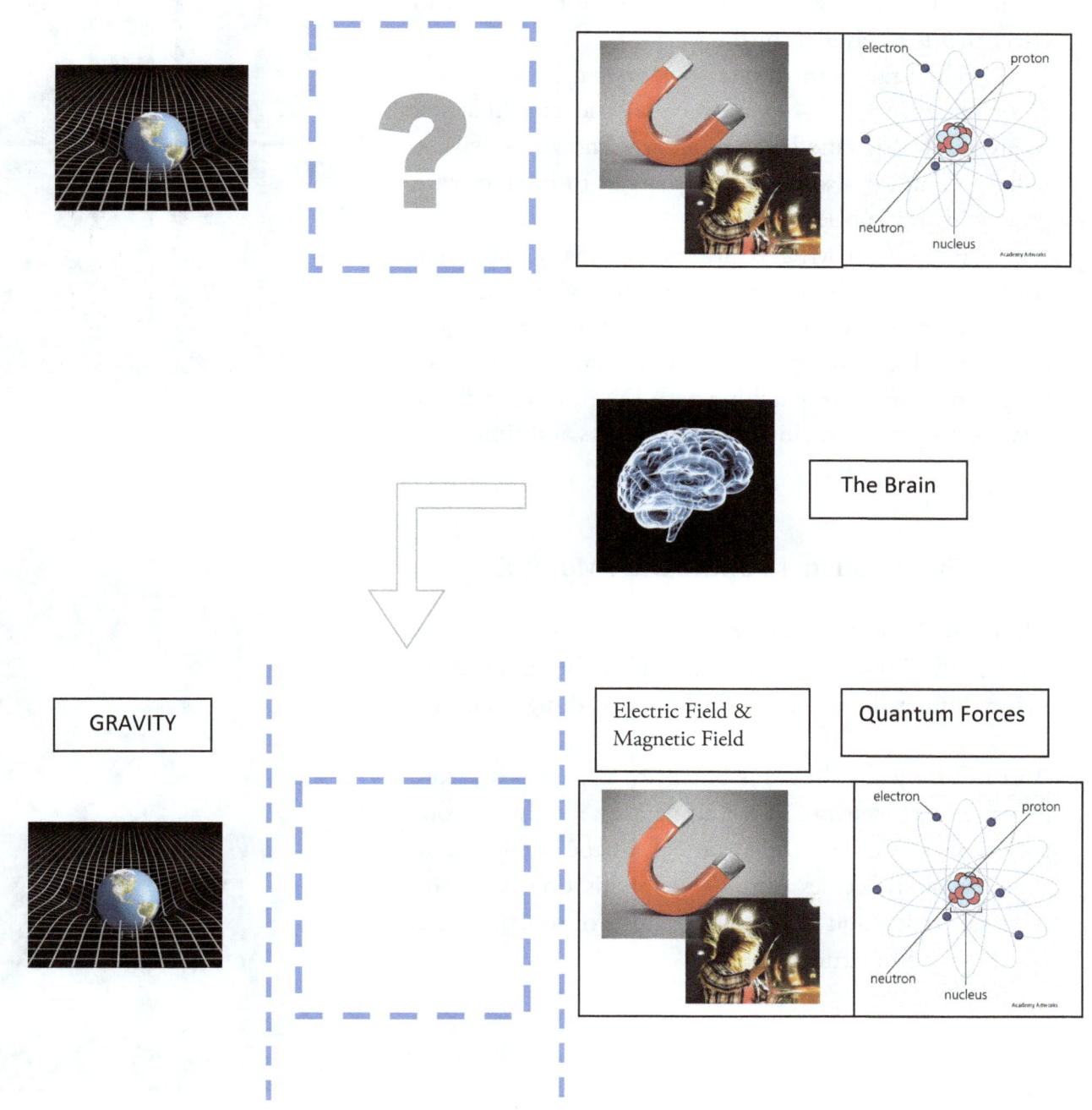

The Uniqueness of the Brain

The brain may be the only known naturally occurring interface between what is known as gravity and the other known fundamental forces.

Unlike other cells of the body, brain cells remain permanent, so they would permanently store any type of effect that is received at birth.

The brain and the nervous system houses a vast array of sensory sub-organs such as the eyes that are receivers of a specific band of energy known as "visible light". Could it also be the only organ to interact with gravity in very subtle ways?

Functional MRI is just one of many tools that confirm that the brain is the center behavior. You have to go where the evidence takes you. We focus on the brain over DNA using simple observation. We do not see personality traits matching with DNA, but we do see personality traits matching with birthdays and times.

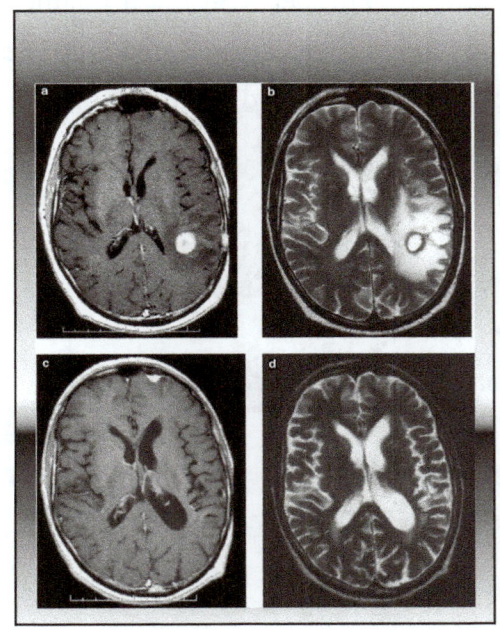

The Uniqueness of the DNA Molecule

Think of how "far-fetched" sounding it is that a UNIQUE single strand of molecules found in every living thing is the blueprint for how that thing is built.

And not only that but this blueprint is passed on from generation to generation. And again it's only this one tiny molecule called DNA, out of all the millions and millions of molecules out there on this one passes on physical traits and determines most of our physical appearance and structure.

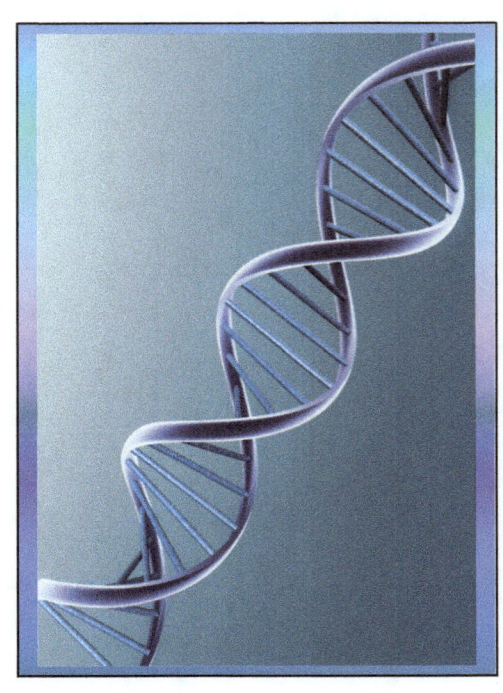

We all have the same basic brain functions; well-known brain anatomy and bio-chemical known as neuro-transmitters. Also the cell functions of axons, synapses are all well known. But we don't have the same memories, or the same behavioral response to the outer world. The focus has to be on the differences, what distinguishes one brain from the next?

On a behavioral level, we do not share the same responses to the same environment, as subtle as these differing responses may be. This is a clear sign that something is different internally. The question is where does this difference lie? Is this difference in the vast inter-connections between brain cells or is it something deeper than the cells, possibly at the molecular level? The brain with its various structures and functions remains one of the greatest mysteries in modern science, it is a physical example of the coming together of the material, molecules and cells and the immaterial, thought and consciousness.

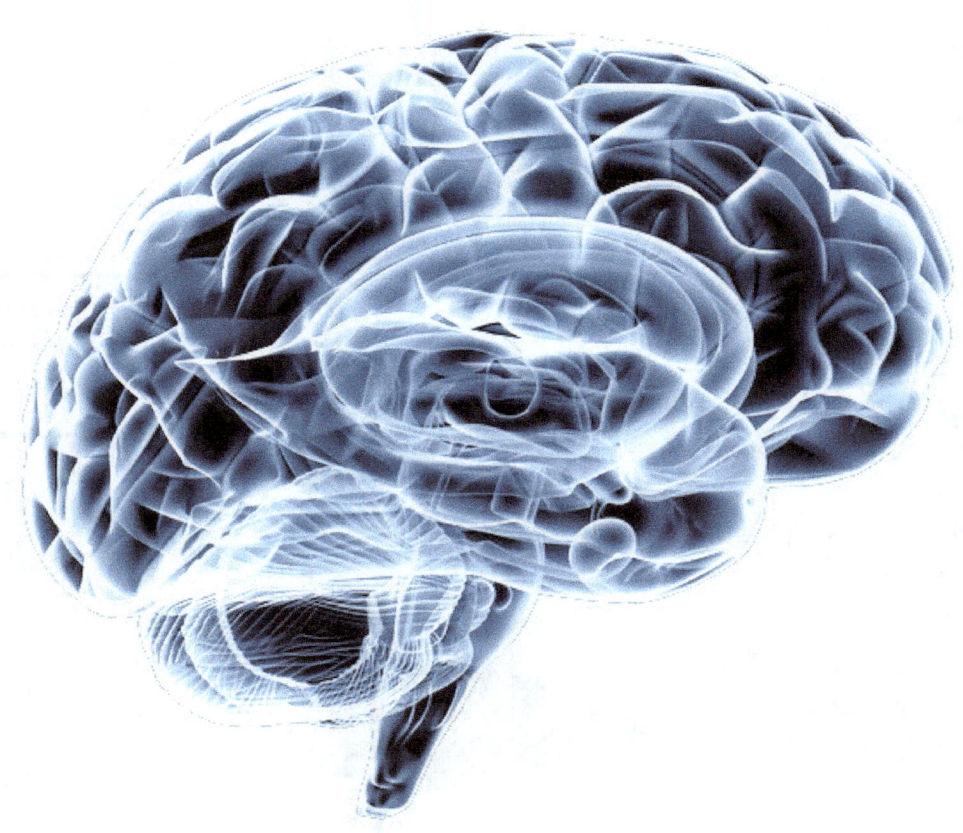

A Related Discovery Confirmed

The long standing mystery of birds observed to possess astounding abilities of navigation across thousands of miles, at times flying at night, flying under cloud cover with no visible stars or landmarks of any kind led some to speculate that there was a biological compass or an ability to sense the Earth's magnetic field. Faced with this mystery, scientists began to search for evidence of magnetically sensitive material within birds, formally establishing a field of research and study dating back to the early 1970's, called Magnetoception. This is a large area of study that deals with the natural ability to sense and respond to magnetic fields, covering a wide variety of animals from bacteria to fish and also includes humans. In the case of birds, a major breakthrough occurred when a duo of scientists confirmed that magnetic fields are sensed within specific brain area of pigeons.

Four Main Elements of the Discovery Process

- **Observed Behavior – Unexplained Bird Navigation**
- **Biological Compass Theory – Earth's Magnetic Field always present**
- **Experiments & Research – Identified Sensory Organ**
- **Experiments & Research – Storage Ability within the Brain**

Initially, it was thought that the beak contained either the active magnetic material known as magnetite (Fe_3O_4), or possibly a bacteria capable of sensing magnetic fields. However, the most recent findings determined the active substance to be **ferrihydrite** ($5Fe_2O_3 \cdot 9H_2O$) believed to be located in the inner ear region of the pigeon.

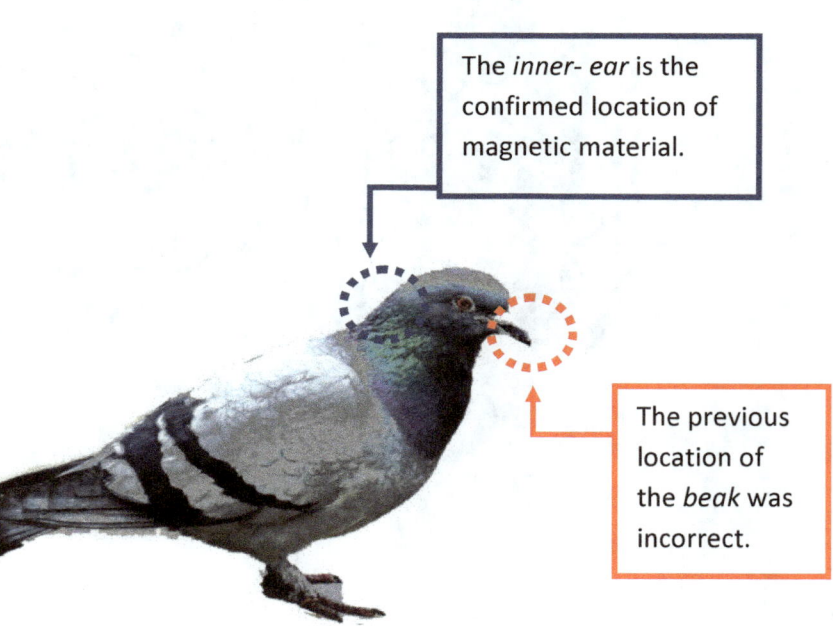

The *inner-ear* is the confirmed location of magnetic material.

The previous location of the *beak* was incorrect.

Figure Navigation by Brain

The Essential Brain

The research* focused on brain cells located in the brain stem. The researchers followed a link to the inner ear, the source of the sensory stimuli. Their research was able to show that specific brain cells responded when a bird lined up with magnetic field direction and also responded to field strength. The research is extremely enlightening. However, they were not able to pin-point exactly how the magnetic field was being sensed. But the inner ear appears to be the sensory center for this magnetic reception as opposed to the beak. The presence of magnetite has also been found in humans, and experiments have been performed to determine if humans have some ability to navigate using this sense as well. This is an important example of a natural field that surrounds the earth interacting with the brain of an animal, supported by hard science. Therefore, it shouldn't be that much of a stretch to suggest and explore the possibility that the human brain can also interact with the earth's gravity.

Bird in flight depicted as sensing magnetic direction, just as ferro-magnetic materials naturally lineup and orient themselves north to south just like a compass needle.

The magnetic field of the Earth, although it is not extremely strong, is accessible almost anywhere anytime.

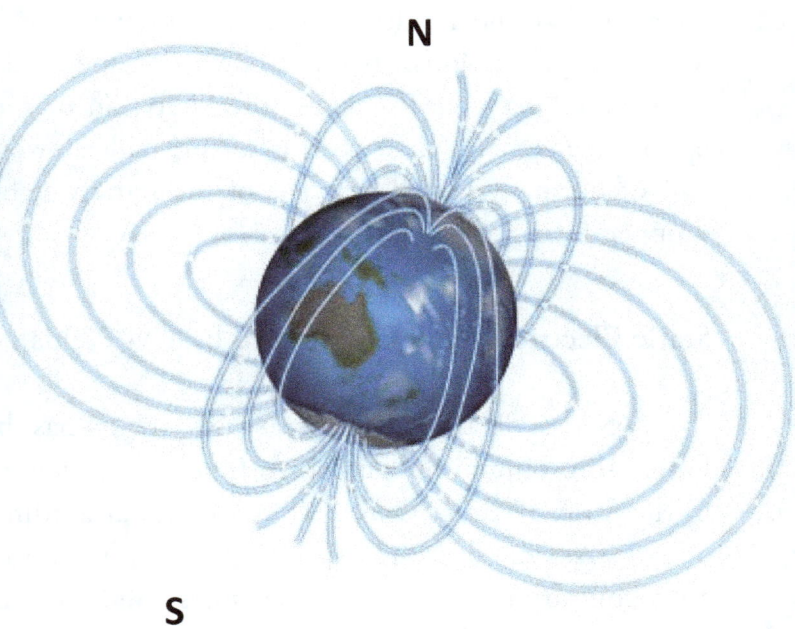

Magnetic reception in organisms and animals is called MAGNETOCEPTION.

*Source: Le-Qing Wu, J. David Dickman
Neural Correlates of a Magnetic Sense
Journal Science 25 May 2012: 1054-1057

Possible Interfaces

"Gravity" & the Human Brain

There are several possible ways in which gravity could affect the brain. The following are some examples of forces and devices that exhibit interaction with gravity and the Earth's spin. All of these examples are artificially occurring, the one place some form of these may naturally occur is the brain. The first two examples appear to interact with the field component of gravity and the result is a reduction in the pull or force component of gravity.

Opposing Magnetic Fields (coupled)

Inventor and research engineer Boyd Bushman Sr. of Lockheed Martin Propulsion revealed, according to recorded demonstration, that by forcing two opposing magnet sides together that the resulting field would interact differently with the gravitational field. The experiment claims to reduce the "mass effect" or pull of gravity. So that a pair of opposing magnets (eg. 2 norths facing) falls slower than a pair of normally joined magnets of the same weight, when dropped from a roof top.

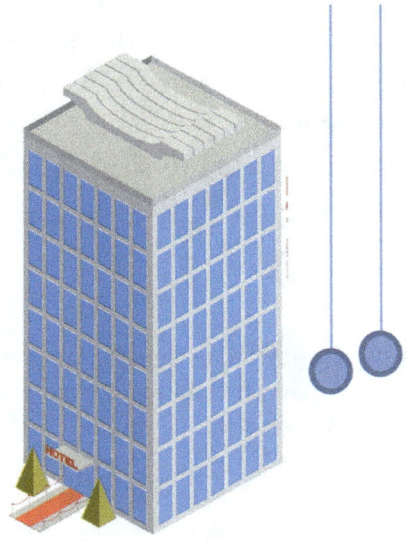

Sphere of 2 opposing magnet's falling slower than sphere of 2 normally joined magnets weighing the same.

Static Electrical Field Overlapped by EM Fields

Some of the work of Nikola Tesla has been reproduced and demonstrated by John Hutchison. Hutchison has shown that by producing static fields of electricity then introducing EM in the form of radio waves among others extraordinary reactions take place. Many of Hutchison's demonstrations show reactions that interface with gravity, causing objects to levitate, almost appearing to cancel out gravity at times. Videos of his demonstrations are all over the internet and have been featured in a number television shows. Of course, mainstream scientists publicly call his videos and experiments hoaxes, but never has a scientist challenged the idea of overlapping these fields of energy.

The following forces and devices are more related to orientation and spin of the Earth than directly to gravity. Is it possible that we have the sensory ability to detect subtle movements of our planet?

Coriolis Force

The direction water spins going down the drain varies depending what side of the planet you are on, this is called Coriolis force. This is a force due to gravity. Water goes down clockwise in the northern hemisphere and counter clockwise in the southern hemisphere.

Figure 75a

Figure 75b

Celt (pronounced: selt)

Also called a rattleback, these objects have been found in Ancient Kemetic / Egyptian archeological sites. They also have a Celtic origin, now used as a toy to show spinning reversal. These "toys" only spin continuously in one direction. If you attempt to spin a celt in the opposite direction it will simple stop after a few spins.

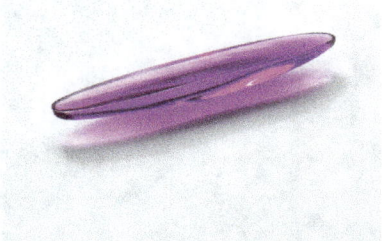

Figure 75c

Foucault Pendulum

By special design the motion of the swinging pendulum actual "detects" the spin of the earth as the targets below the pendulum are struck when the earth moves slightly out of sync with each pendulum swing.

Figure 75d

THE CAPABLE BRAIN

If a special combination of static electric fields and electromagnetic fields (EM for short) are what is needed to interface with gravity then the human brain has that times a billion! The human brain having billions of neurons means that EM is extremely abundant in our brains, this bio-electro-chemical quality makes it unlike any other "object" known to exist. This also means the necessary pieces for a possible interface with gravity are present in the brain. The brain's micro electric currents produce EM, that are one half magnetic field. Could any of these be set up in an opposing fashion or in an overlapping fashion to interface with gravity? And are there specialized brain centers that perform the amazing feat of interacting with gravity, to sense planetary movement?

FACT #1
The Human Brain is the largest most complex assembly of naturally occurring biological static electric fields and electric currents.

FACT #2
Electric currents produce **EM – electromagnetic** fields / waves.

THEORY
If both a gravity interface and human memory depend on overlapping fields and EM, then brain has the most of both in one place!

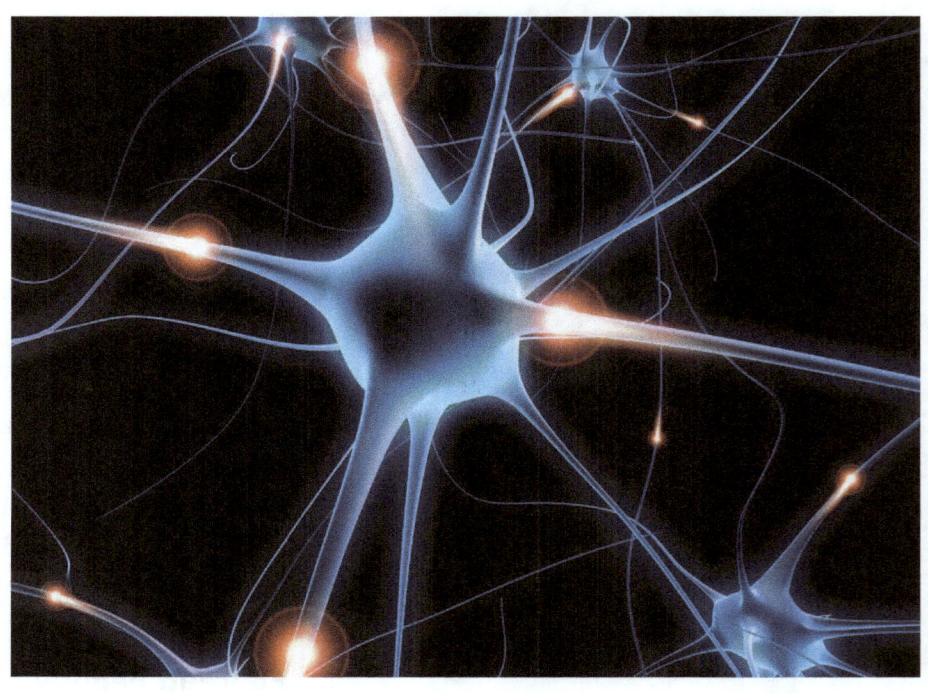

We have our theory, and we have a much better idea on what to focus our search on. So now we can concentrate on the many invisible fields in the brain rather than just the cells and molecules. There is a strong case for the idea that the charges, currents, and fields in the brain are not mere byproducts, but are equally essential to brain function as the cells and compounds that make up the brain. Obviously these fields, are extremely small, and exotic overlapping fields may occur naturally inside the brain. These special overlapping fields may prove to be extremely difficult to find. But the small size and complexity of our mysterious brain areas are not a problem for the brain's function. Magnetoception, like the navigation of a pigeon, has already worked out that tiny regions of the brain can interact with fields of minimal strength, with molecules of magnetite on the nano scale in size, much smaller than a brain cell itself.

THE ESSENTIAL BRAIN 77

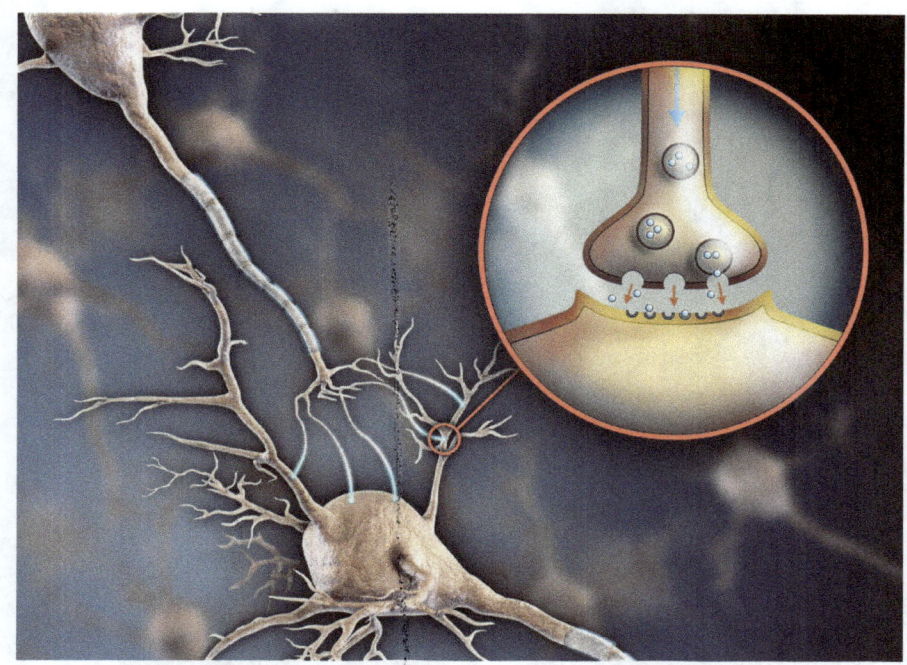

Figure 77a

The basic building blocks of our bodies are cells. It may surprise you to know that almost all of your cells have a static charge. The very cell wall itself has a dual layer with a positive charge on one side and a negative charge on the other – it's called the bi-lipid layer.

But our brain cells are even more extraordinary, they actually build up charges and fire them out to other brain cells –this is called a synapse.

One thing that may have been overlooked is *Vanic radiation*. This is a special type of EM or field produced by a current. It acts like an antenna sending out EM, only in this case it may be possibly helping to create an "antennae for gravity".

Let us go back to the chemistry of the brain for a moment. The human brain is a huge factory, filled with electro-chemical compounds from simple ions to neurotransmitters. All of these constantly flowing and reacting with cells. Wherever there is moving charge there is current. Where there is current there is EM. So on every level, we see the presence of fields, now it is time we take them into consideration.

Are these vast arrangements of fields sensitive to the position of the planet and the overlapping gravitational fields upon the earth? Is this what creates our first memory just like overlapping light creates a hologram on glass, which also happens to be subject to static fields? We need more clues, let's continue to explore, and there is no better place to start than with birth.

Figure 77b

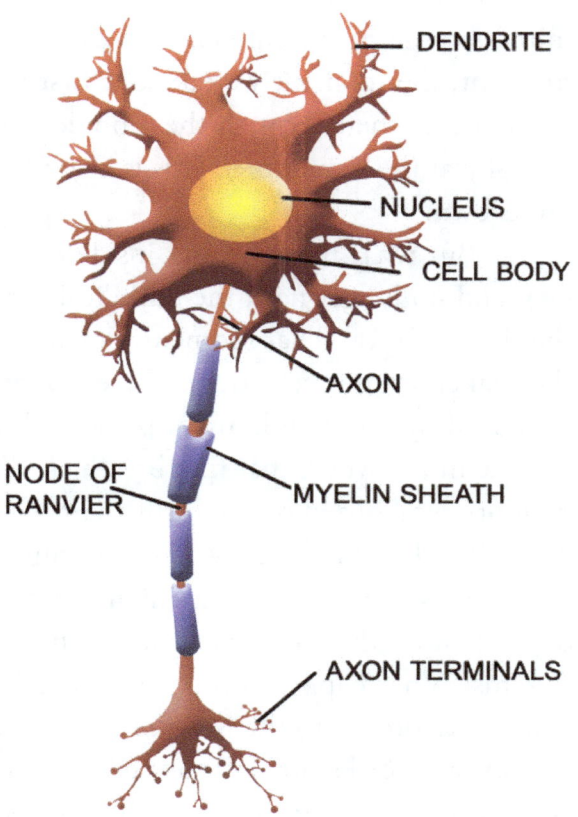

THE BIRTH MOMENT

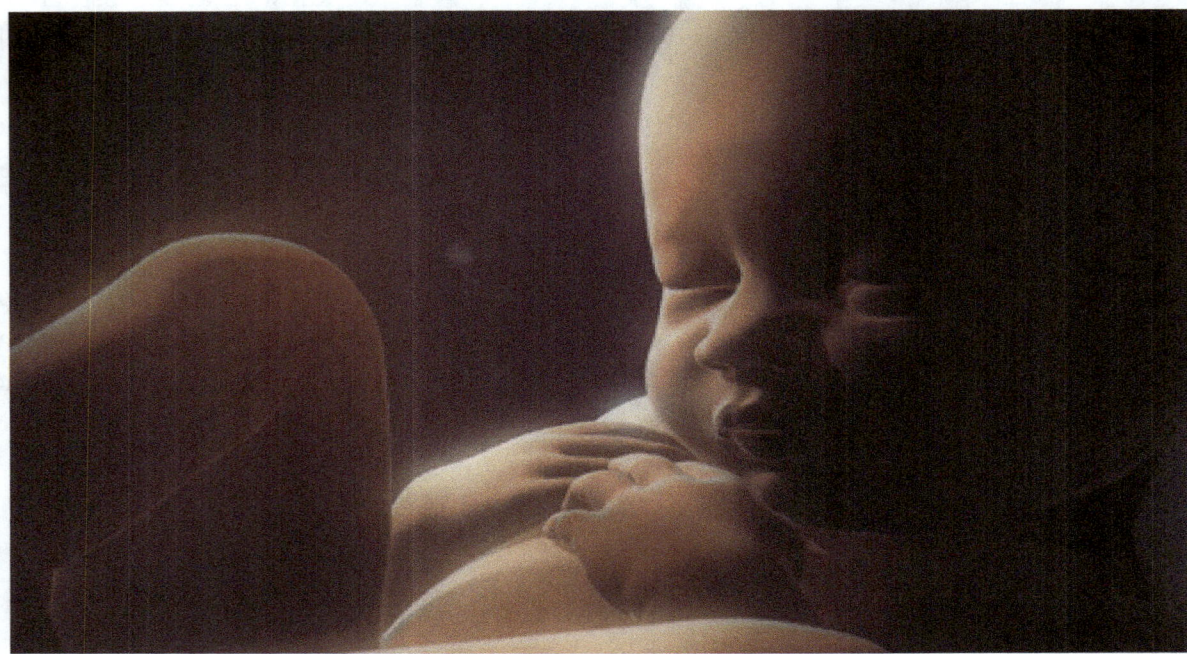

The precious and special event we call birth is the time most likely to account for the initial storage of the influence imparted on us by planetary forces. It may all come down to a moment, the moment we take our first breath. Up until that moment, we are receiving oxygen from our mothers through the umbilical cord. Many times the first cry and the first breath come virtually at the same time. This is a good sign the doctors know – this baby is alive and breathing on its own.

But what is happening inside of our little baby brains at the moment we breathe on our own and no longer need the umbilical cord? That very moment is described in the birth chart as the time of birth. That moment, down to the second, marks the storing of the of the positions of the planets at the very time. Like a snapshot of the heavens, or a recording of a unique tone produced by the symphony of planets. From our nine months in the womb to our first minutes outside in the world that first breath marks the Birth Moment and the storing of the first cosmic memory symbolized by the birth chart.

So then biologically, this one-time event of birth, must be major as well. Certainly some major transitions occur. The infant goes from "life" within a life, the environment of the womb, dependent on the mother for all sustenance. It's also a huge event for the respiratory and nervous systems. What happens on a neurological level at that time of the first independent breath being taken? According to the Kenemonic Model, this is the time that the "music" or tune of planetary motion is recorded or statically stored in our brains in a specific brain center. This "saved" effect is a form of memory storage. The idea is that our brains, even before birth, have the ability to "receive" the planets effect, but at this special moment a type of permanent memory is formed.

The idea is consistent with other factors that shape us from the moment we are conceived from the combining of the egg and sperm. From the moment we are conceived the environment starts to affect us. The health of our mother, the food she eats, and more factors than we can list here affect the unborn child. The point is, that the planets are one of these major factors, that will go on to influence and shape our basic personality traits.

The Storage of this solar system effect is a form of memory, but an involuntary memory. Because we never consciously sense the planetary force, but yet we can store and react to the effects.

THE ORIGINAL MEMORY

The speed of a synapse even millions of them have enough time to form and imprint a subconscious memory of this event - the first memory. But it is even more than a memory, its part of our essence.

In the birth chart, the rising sign changes to a different degree in an average of 2 minutes. The Moon sign changes by one degree in about 2 hours. So two of the top three factors of the birth chart change extremely fast.

We are truly children of the solar system, as the positions of the planets store themselves as personality traits that we will have for life.

The rare cases of head trauma that lead not only to memory loss, but personality loss also support this idea of a birth moment that crystallizes our personality traits. So barring some type of rare head injury, it seems this is the time we get our personality traits for life.

The Initial Effect

The theory is that, there is a portion of the brain or specific brain centers that are capable of sensing the forces of planetary interactions and permanently storing them. The musical analogy fits as well. This solar system effect, gives us our base personality and also serves as the basis for compatibility with others. If you think of it as tunes within a tune, then you can see exactly how people would be harmonious with each other across a wide spectrum of possible levels almost instantly when they first meet. And also how certain combinations of people are not harmonious.

There is a constancy of the brain, although the brain continues to grow after birth it does not replace the core brain cells. This is one of the reasons the brain is capable of storing our basic personality traits.

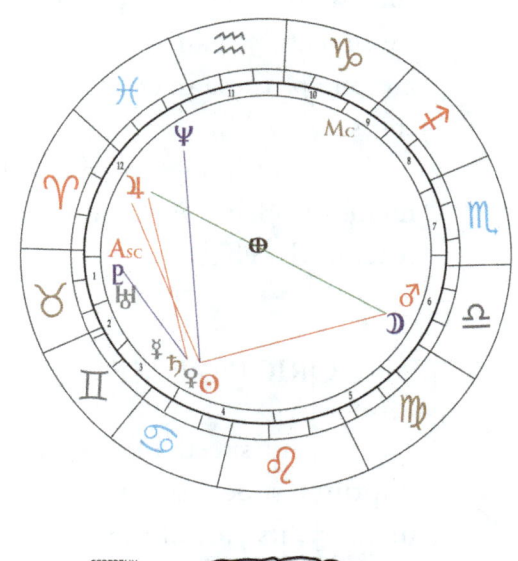

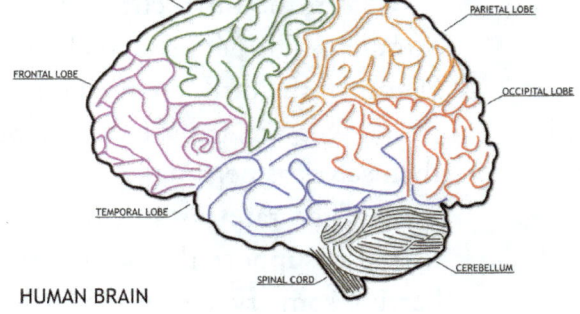

HUMAN BRAIN

The result of the solar system effect at the birth moment gives us the brain's equivalent of an individual's unique DNA. The major difference being that these personality traits are not inherited from parents, they are imprinted onto our nervous system at the moment of birth.

This initial effect, of our personality traits endowed by the solar system does not determine our future. This is just one of several strong factors that include free will that influence our lives.

The Continuous Effect

The solar system effect on the human brain is not just a one-time event, it is always occurring. Just as the planets do not stop moving, neither does our ability to interact with this force. The continuous movement may affect a separate brain center than the center that stores the static configuration. In a complete picture, the brain continuously reacts to changing outside forces. The constant motion of the solar system is a secondary effect, but can have a wide variety of influences in a range of life areas.

In astrology, this is called a transit horoscope, it can be done for a day, even for an entire year. Although a horoscope covers all life areas, in my experience the most requested areas are relationships and money or career.

Similar to the compatibility or incompatibility between the configurations of two individuals there is a constant pairing between the traits at birth and the changing position of the planets. The practice of creating individual horoscopes by matching up the entire birth chart to the current planetary positions, is the called transits in astrology, and is of incredible value to many people who find it a helpful assistance in maximizing their lives.

For example, the classic shared influence of Venus aligning in the horoscopes of two people meeting providing a romantic catalyst. If Venus is aligning favorably with either person's Uranus or Rising sign in real-time this can result in serious romantic sparks and mutual receptivity. Not unlike a favorable weather forecast for sunny skies, shared conditions that may lend itself to many enjoyable outdoor activities.

The idea is that there are brain centers that sense these temporary planetary alignments. Again, this is not a 100% prediction of events, there are simply too many factors that have to line up. The power is that everyone is affected, even though we are affected differently, some have common reactions.

The Brain & Memory

Huge projects are underway in the fields of Neuroscience to find the physical nature of memory storage and to map the billions of neuronal connections. Both will yield the evidence for individual variance and configuration. And it is central to the theory of Kenemonics that this will also herald the correlation of these configurations to the positions of the celestial bodies at the time of birth.

We can certainly scan the brain with scans like the well know MRI and PET scan, these allow for imaging brain areas that are active given certain activity and stimuli. The more difficult challenge is at the cellular level and beyond that the molecular level. The sheer magnitude of the interconnected brain cells and the huge amount of specialized functions for both the cells and the brain's bio-chemicals including the neuro-transmitters make the task extremely resource consuming.

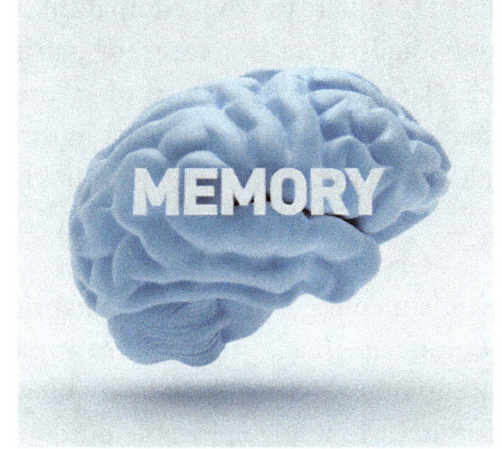

Logically, we also have to consider the possibility that memory is not a complete mystery, but a guarded secret. There are levels of secrecy, even inside corporations, let alone "black projects" run by government agencies.

At this point, we can only speculate on this topic. But it is extremely important to consider the possibility that, in large part, human memory may have already been figured out by a clandestine few.

Mainstream science has its own memory theories. There are widely accepted areas of the brain like the Hippocampus that are considered to be the center of memory. There is the concept of "plasticity". The the brain's changeability, in making new neuron connections, when there are new actions this helps form new memories . There is also the idea that the complex connections of neurons may be a part of memories as well. But that doesn't seem likely. It is more likely that the neurons / brain cell connections are formed to support and reinforce memory, but are not actual memories themselves. If memory was only structure and links of neurons then mainstream science would have figured it all out by now - but they haven't.

So then, human memory, including the birth moment is likely stored as energy. It is in the form of complex electrical fields supported by cell structure and electrical currents throughout that structure, not as molecules. Looking at all of the possibilities the true essence of memory is, not neurons forming links or molecules arranged in a complex way, but energy – energy fields stored. However, the changing structure of neuron links is also likely linked with memory storage and formation.

Brain storage seems to be holographic. But, is it holographic in a similar way to optical holography? Certainly brain storage is not optical. But does the principle method of overlapping energy in the optic case, operate within the brain? Can the same holographic storage be achieved with EM fields that are produced by the tremendous amounts of electrical activity within the brain? Given that optical light is merely a segment of EM energy, and that the brain has to be full of EM because it is full of currents, then the possibility of some form of holography may be there.

With regular photo image holograms, there is an overlapping of light patterns and a target. The resulting optical plate holds the image, even fragments contain the entire original image. The brain exhibits this quality as well, there are many recorded examples of brain injury in which large portions of the brain become damaged yet the brain is still able to function by utilizing other areas. Storage technology based on holographic science can store more information than any other technology known by size.

So in the case of a hologram what is being stored? Is memory the energy stored? Start with 1 sheet of glass. Imagine every part of the glass is basically the same. Now cut the glass in half. The 2 pieces are still the same. One piece we are going to expose to hologram and store an image in it. The hologram becomes visible in the glass. But did the glass change? Something is there in the 2nd piece of glass that is missing from the first. The second piece of glass has trapped the holographic image or energy – or stored it as a memory.

The Brain & Memory

Personality traits stored at birth is a process is very similar to sensory memory recall, only there is no sensory information stored such as pictures, smell or tastes. Memory and the mechanics of Kenemonics work in a similar manner. Storage & Recall. Reference & Resultant action. In the case of personality the process is dependent on "structural recall" for a portion of behavior and response. But still it is still possible for the method of storage to be the same.

If you lose your memory, certainly your actions will change. We all act in certain ways based on memory alone. We know where we live, and who are friends and family are. So just like those basic memories, personality traits are also constantly being accessed and referred to, and both are a basis for actions and reactions. This is shown clearly in certain cases of severe brain injury. Not only can you lose your memories through extreme brain injury, but personality can also be lost or altered dramatically. Let's go further, and look at some real life examples.

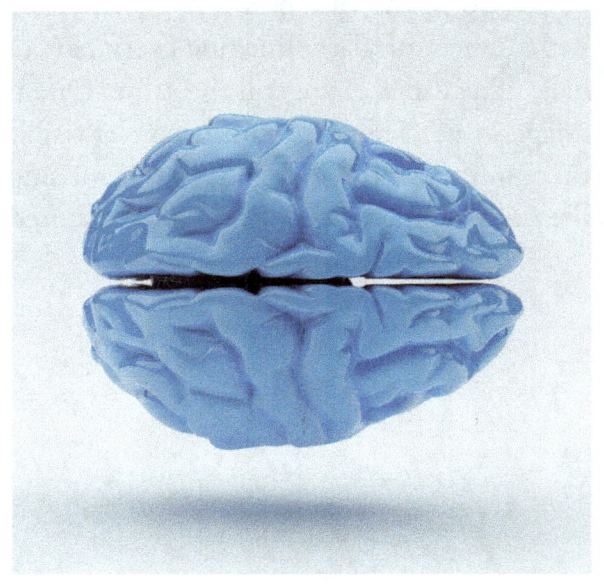

Everyday most of us leave home and return. It's a simple act. But what if we lost our memory while we were away from home? That would make it difficult to get back home. So let's look at Case A, a simple but powerful example of how memory works to make life function.

First, there is the memory of your house's location and appearance. On your way home, you have to retrieve that memory. Next, you act on that memory. And finally, you arrive home.

Now let's take Case B. Starting at the moment of birth, the solar system's unique position at that time is stored in the brain. This effect gives the core personality traits called kenes. These kenes are stored inside the brain similar to how memory is stored.

From the birth chart, we have Virgo Sun, Mercury in the 6th House, and Uranus conjunct Midheaven (start of the House that rules the career). The actual birthday is September, 16 1984. So there are several kenes that show a possible medical and scientific interest. The actual result in this case, is multiple degrees in science and a career in the medical field. This is more than just the simple repeat actions of Case A, but the idea is the same.

THE ESSENTIAL BRAIN 85

CASE A

Information Stimuli

Reference Storage

Memory Stored

My house, location appearance

Action / Results

CASE B

Information Stimuli

Solar System Effect

| Birth moment | brain |

Stored & Referenced

KENES

Personality traits
*Specialized Memory

| Birth Chart | brain |

Action / Results

Career (Possibilities shown in Birth Chart)

Brain Centers & Neurons

We are looking for two things. First, what part of the brain is responsive to the solar system effect. And second, where personality traits are stored in the brain. Both may be located in and around the same brain area. Possibly even part of the same system. That is, the powerful system known as the limbic system. Can this be the area receiving and reacting to fluctuations of planetary alignments, storing them at birth, and continuing to resonant with changing alignments without us ever consciously perceiving it?

If the limbic system is the correct area, we have to design experiments to test and verify it. The basic methods of observation may not work. The limbic system is deep inside the core of the brain and has many complex parts. There could be a vast array of EM fields acting as sensors in the brain only interacting with gravity or serving multiple purposes while invisibly interfacing with gravity.

RESEARCH GOALS
1 Individual "unique" configurations will be discovered.
2 Brain similarities and dissimilarities will be discovered between individuals.
3 Next theses configurations will be correlated to planetary alignments at the time of birth, and personality traits should correspond strongly to the birth chart.
The first 2 have already happened in the field of genetics, we just need to repeat the process with the brain and personality traits..

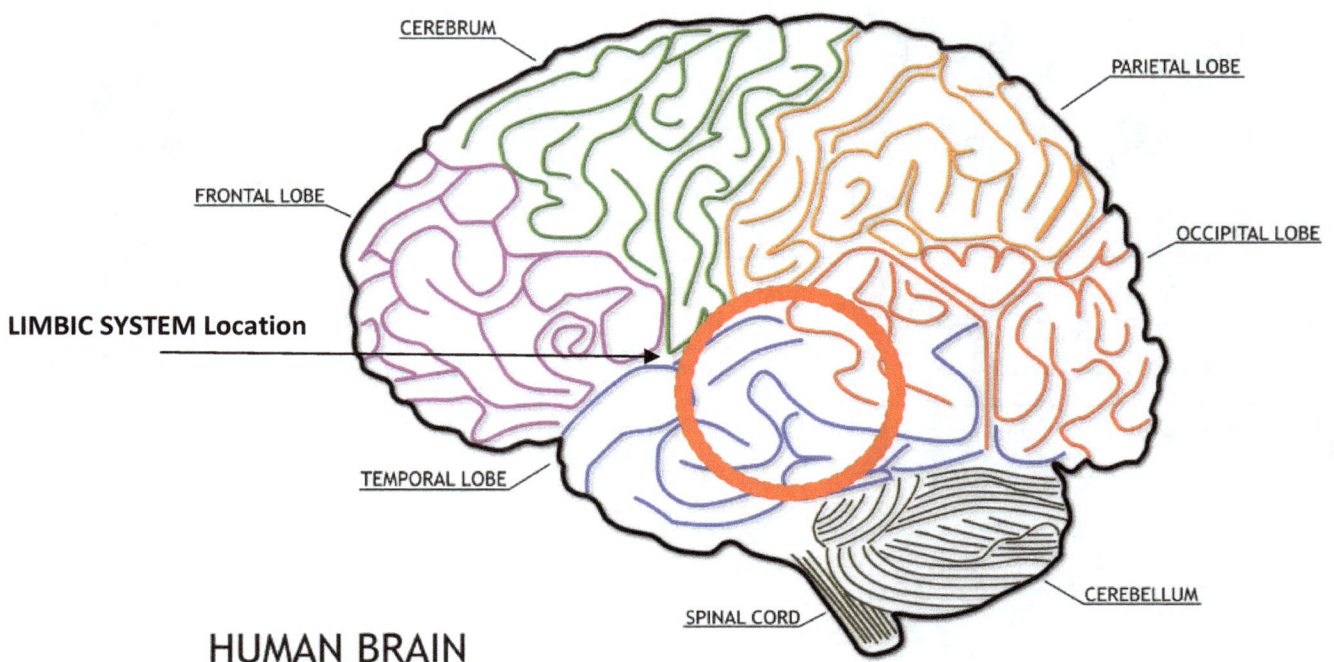

HUMAN BRAIN

Hidden Brain Structure

The well-known sense of sight can be explained in scientific terms, perhaps in more fantastic terms. Incredibly, the brain has sense organs that sense a very special portion of electromagnetic radiation called visible light. It collects that light and forms images in our brains to enable us to interact with our surrounding environment. Further, these images are "often if not always" stored as memories within our brains. So just one of the many attributes of the brain is its ability to actually sense and record one of the fundamental forces of the universe, that being electromagnetism.

It's not so much of a stretch to conceive the brain being able to sense another fundamental force of nature, in this case gravity. So the same way we can take in reflected light and form images and memories, it is highly probably our brains are doing the same with the gravity of the sun, moon and planets. The force that is the basis for planetary interaction and one of the basic forces here on the earth. The only difference is that we are not consciously aware of this sense and its influence as we are with our sense of sight.

A large part of the problem is that we just don't see, taste, feel, or hear some of the most important things that are happening to us. But like the example of Vitamin D production from sunlight on page 66, we can see the effect if we look deeper. Astrology's claims of influence on personality are showing up in human behavior - highly specific unexplained preferences such as physical attraction.

"It Is Entirely Possible That Behind The Perception Of Our Senses Worlds Are Hidden Of Which We Are Unaware."

Albert Einstein

The basic building blocks of the brain are the cells, axon, synapse, receptors, dendrites. Can these for more complex circuits? Specialized like the optic nerve and auditory nerve yet be completely subconscious to us, unlike the 5 senses.

Where exactly in the brain could this take place? A sensory area that is sensitive to gravity would not need to be exposed externally on the body, like the skin, eyes and ears. For reasons that should be obvious, gravity cannot be blocked, so any brain area capable of interacting with this force could be embedded deep in the brain. What we are looking for may be a complex chain of interactions in which an actual physical configuration is only one piece or link in that overall chain. Still the first order of business is to find some physical structure that definitively demonstrates stored memory and some individualized structure within the brain.

THE BRAIN & HUMAN PERSONALITY

It is clear that the brain is the seat of human personality and behavior. For lack of better words, it is the place where thoughts happen. The challenge is to explain how personality comes to be and how it is stored in the brain. Based on our study of the birth chart, personality traits come from the solar system at birth and are stored similar to other memories. We have behavior as evidence of this, but not complete evidence yet.

Serious brain injury also offers some clues. These injuries can change personality, there are even cases of new talents. These brain injury events that alter personality supports the idea that personality is a product of the brain not genes. Obviously, genes are not changing when the head is struck or seriously injured.

If we can identify consistent physical differences in individual brains showing certain personality traits that would be an extreme break through. But even though this may not be exact, it will be more advanced than our current level of knowledge and understanding, a deeper understanding that can hopefully lead to more.

THE GENETIC FACTOR

With the achievements of the Human Genome Project, genetics has become science's "shiny new hammer" and everything looks like a nail. So there has been a mass effort to link everything from behavior to disease to genes. The possibility of a connection with genes may be there. However, in Kenemonics no theories of genetically influenced behavior have been included. This is due to the lack of scientific proof that genes directly affect behavior.

Genes clearly affect physical traits, traits from the parents can clearly be seen in children as well as shared physical traits between siblings and other close genetic relatives. There are genes that affect the brain. But genes that affect the brain and its chemistry are not effecting personality traits, they are causing a condition.

It is apparent that our genes are responsible for "building" and providing the blueprint and plan for our brains, which give us a great range of cognitive abilities. But once the brain is built it is the brain that takes over. But in the final summation, genes are not a direct factor in the making or storing of individual personality traits.

SECTION 4
PUTTING IT ALL TOGETHER

Up to now we have dealt mainly with astrology. First we uncovered the mysteries behind the solar system science of the birth chart that shows us our personality traits given at birth. Then we explored deeper into the incredible solar system effect and the best theory to explain it.

In this section we will put everything together, combining personality traits with the other factors of nature and nurture. This will give us a clear picture for understanding how all of the factors of life come together. In the end, we will have a template that includes all known things that go into making us who and what we are as individuals, taking into account our genes, personality traits and our life experiences.

Harmonics

Most of us know harmony from music. We love our favorite songs, we dance and sing along. All the while we never think, or need to think, about why it sounds so good. Yet there is a reason. There is a principle at work – harmony. Harmonics or harmony is a basic principle which seems to be present everywhere we look, throughout science, nature and life in general.

Harmony basically determines what "things" should be combined together and which should not. Theses matches range from relatively positive to relatively negative, compliments, and clashes. We even use harmony to match the colors of our clothes. And of course we experience harmony in our personal friendships and relationships. Even after all of that, harmony is not always appreciated for its role as a natural part of life.

Harmonics generally start with something in motion. A simple form of motion is vibration - a tiny back and forth movement. With sound, it is the air that is moving. We see colors because light is vibrating at different speeds or frequency. We're not going to get deep into how mainstream science deals with harmonics. We don't really need to because harmony is something we can all see, hear and feel.

We have an individualized response to colors, sound, music, singing, voices, words, even ideas. Harmony is a part of our language. Phrases like; good or bad vibes, being on the same wavelength, that resonates with me, and that struck a chord, are hints that we are experiencing our own form of harmony.

"If You Want To Find The Secrets Of The Universe, Think In Terms Of Energy, Frequency, And Vibration."

Nikola Tesla

Take the word resonate, we experience it in three basic ways. First, when someone says the same thing we have been thinking. Second, when someone says the same thing we have said. And third, someone expresses an idea you hear for the first time and you relate so strongly it inspires you into action.

See figure 91a to the right, the famous example of resonance matching of sound waves with glass.

Harmonics

Figure 91a Scale of Sound - Human Hearing Range

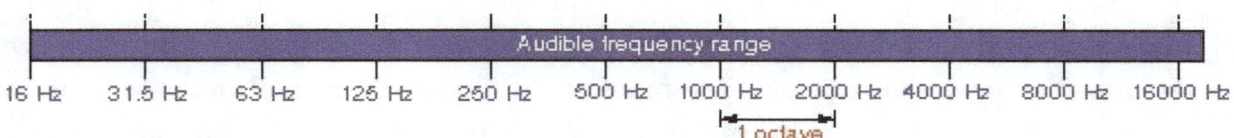

Figure 91b Visible Light Spectrum - Human Sight Range

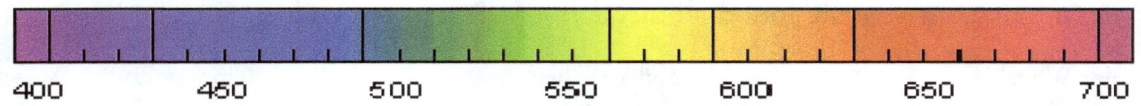

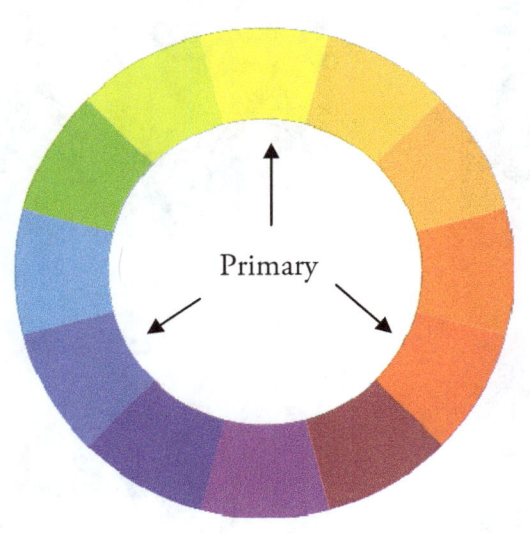

Tertiary Color Wheel
12 Colors from 3 Primary Colors

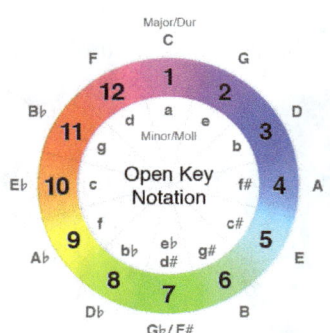

Sound has harmonies and matching musical notes, this is similar to how colors match. With visible light we can see the linear spectrum can be shown as the circular color wheel. In the wheel the relationships of the colors to each other become more clear, this is known as color theory.

We can see parallels between systems, just as colors have natural complements at 180° opposites across the color wheel. The same exists with the Zodiac wheel, at 180° opposite from each sign is its complementary element; fire to air and earth to water, described in detail on page 32. Now we can begin to see the zodiac within the larger context of naturally occurring effects of vibration and energy.

Harmony has a structure that determines how things relate. We can see this visually with the help of shapes. The circle is one of the shapes that clearly shows harmonic relationships. The zodiac wheel is a clear example of this in the area of human life and personality.

Astrology deals specifically with the harmony of the circle, divided into 12 equal parts, and makes it meaningful to our everyday lives.

Harmonics of the Signs & Planets

Below we have the color wheel, the common zodiac wheel and the Chinese zodiac wheel. On the second row we see one of the most important harmonics in a circular arrangement. For the color wheel we have the primary colors, the colors that give us all the other colors. In both of the zodiacs we see the first of 4 compatible groups, all 4 spaces or 120 degrees away. For both the signs and the planets 90 square is difficult and 120 trine is favorable also the 60 sextile is favorable. The 180 is tricky, it can be either favorable or unfavorable depending on co-factors. When accompanied by squares the opposition is less favorable, and when trines or sextiles are present it is more favorable, either way opposites are always powerful. This comparison simply shows that the zodiac reflects the same principles of harmony that are found all throughout the natural world.

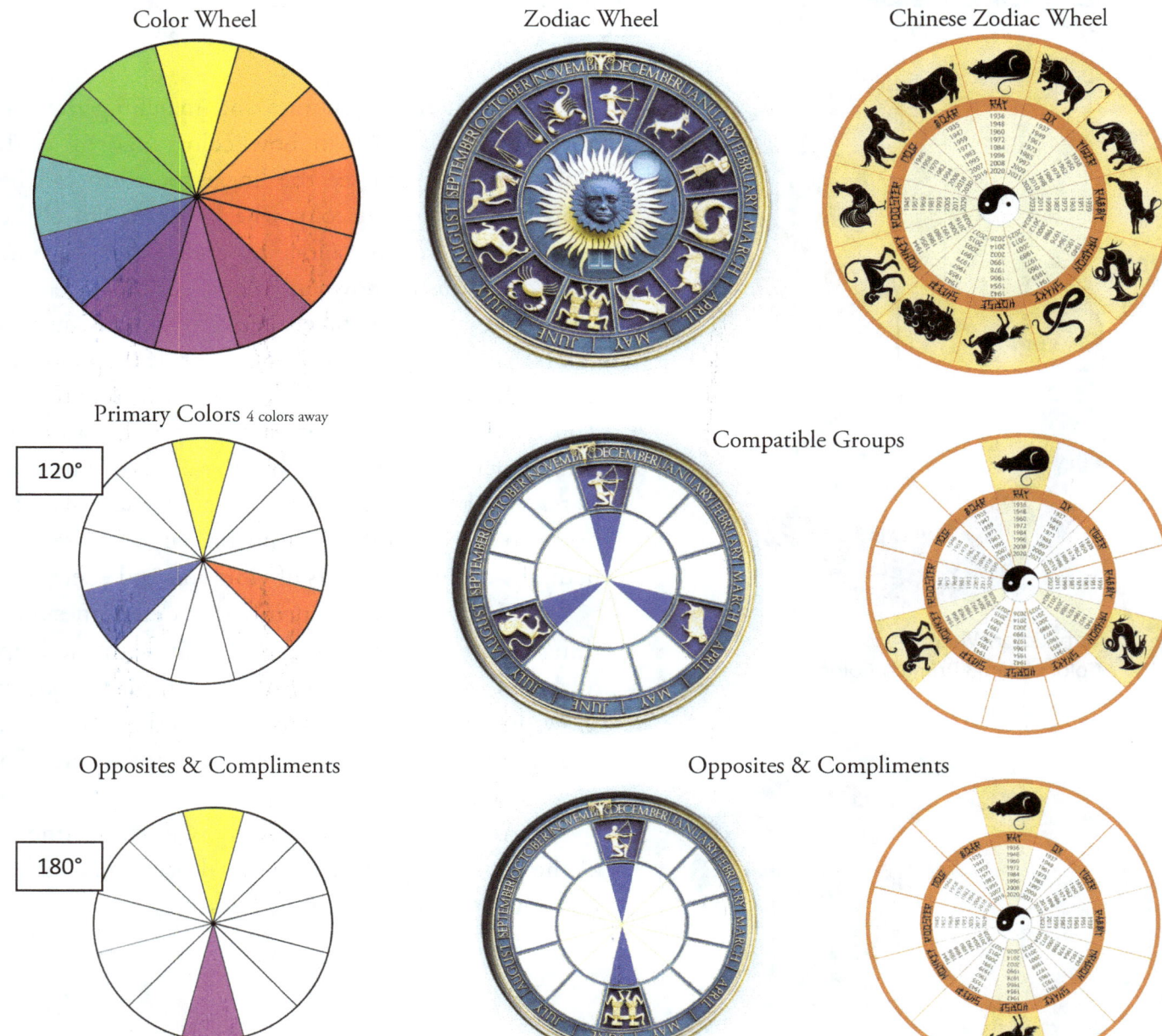

Harmonics of the Signs & Planets

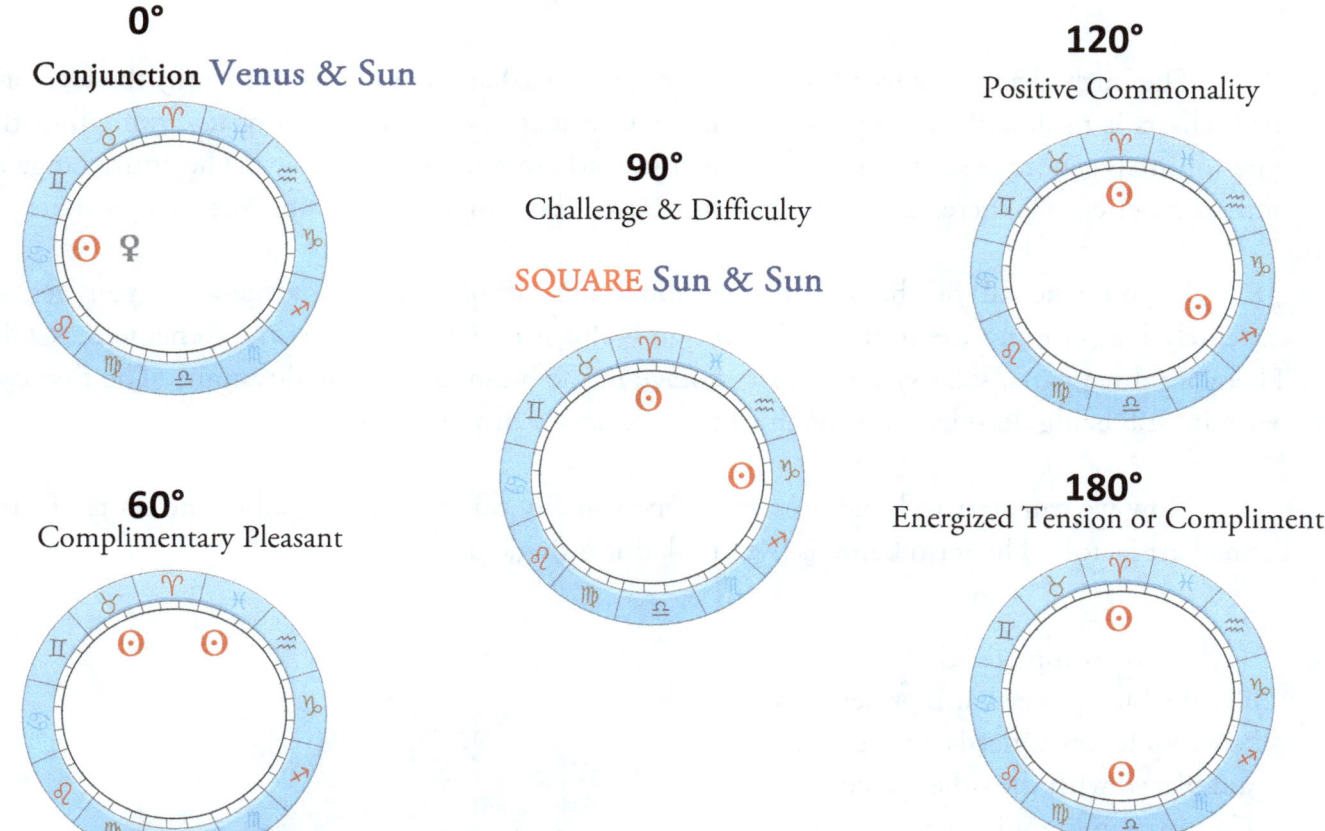

The circle works well for describing harmonies within astrology (some astrology systems don't use the circle). The 0°, or the conjunction, is very powerful especially between the planets of two people. Many conjunctions are classic matches, such as Sun and Venus, that can be powerful factors in relationships and friendships. This combination is found in many marriage partnerships.

There is a degree of accuracy involved, the angles must be within a certain closeness, but not exact. If a combination is within a few degrees the effect is still valid. There are more angles used in astrology, but the five examples above are the most common.

Why not just divide the circle into four 45° quarters? Both 30° and 45° work as divisions of increments for basic geometric shape, circles, squares and triangles. It seems intuitive, but as it turns out 45° doesn't allow for the harmonies to be expressed where 30° does. Twelve divisions come from a natural relationship between geometry and harmonics.

How this was figured out is somewhat unknown, but multiples of 30 and 3 are everywhere. For example, The 30 60 90 triangle, 360° from 12 30° divisions, the 12 month year, 12 inches in the foot. The 12 hour clock, 60 minutes, 60 seconds. The numbers seem to be a reflection of something basic in nature.

Missing Piece From The Puzzle Of Life

The birth chart gives us a list of specific traits in 10 areas of life. In principle, the idea of a birth charts is to describe a range of human traits, a matrix of traits for all of us, from all of the types of intelligence we see to different sexualities and emotional dispositions. The entire range of human experience is there. Like a code of life, inside it is a matrix of possible personality traits.

If you take all of the possible combinations of planetary movements – you get an extremely long list and the sources of each trait – this is one essential part the Kenemonic Code. There are three parts, solar system effect, storage in the brain and personality traits. The first two parts are still being decoded, but the third part we have a firm grasp on.

Now we can start to build a matrix of personality traits, by listing all of the traits of each birth chart factor. The term **kene**, is used to define a single personality trait.

To complete the puzzle of life, the blue piece that represents the kenemonic code needs to be solved and placed with the other pieces.
The surrounding pieces represent the known parts of life, environment, culture, early childhood, and genetics.

This model is not claiming to be predictive or deterministic it is only meant to show that personality traits draw you to certain interests and compel you toward specific tendencies. One of the most powerful personality traits is intelligence. According to the matrix of traits taken from astrology there are eight types of intelligence. Let's take photographic memory, a trait highly related to intelligence if not considered a form of it. It has 3 chart position sources the 3rd is compound. Moon conjunct Mercury, Mercury sextile Neptune, and a compound trait direct from a chart Mercury sextile a Jupiter Uranus conjunct.

Nearly every physical trait can be traced to genes. And virtually every personality trait can be traced to kenes from the birth chart. Every personality expression can be traced to kenes and environment. There is a small area where genes can indirectly "touch" personality, such as genes that affect brain structure or function. For example, down syndrome, or conditions that affect serotonin or other brain chemistry.

Kenemonic Code

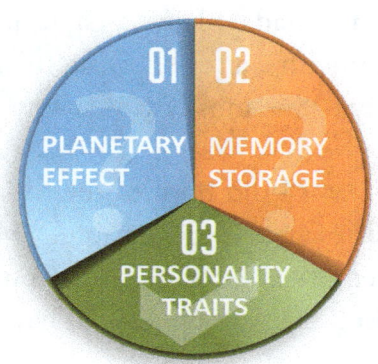

HOW COMPLEX ARE THE PUZZLE PIECES?

10 "PLANETS" used by Astrology (there are more objects)
702 possible birth chart positions, each produces many traits
Your planet positions combination is unique for at least 4,320,000 years. Thousands of Kenes (personality traits), probable ceiling of 5000-7000 kenes – near unlimited combinations.

GENES 4 repeating molecules
23,000 Genes
Thousands or Millions of Trait variations
Unique Combinations – virtually unlimited

BRAIN billions of connections
Thousands of trillions of possible combinations.

The Power of Comparison

As we get closer to putting all of the pieces of life together, it is important to show how to arrive at good conclusions. Also we want to avoid making common mistakes in our thinking. For example thinking that if two people have similar genes, they will have similar personality traits.

In the first example on page 97, we have identical twin sisters. Clearly, they share almost every physical trait. And anyone that knows any identical twins knows that often times twins have similar personalities. But sometimes they don't. And whether or not twins have the same personality traits is not so easy to figure out. Because you cannot tell by simply looking. You would have to get to know each twin.

If we use twin's birth charts as a tool we can often see why some identical twins have very different personalities. Of course twins have the same sun sign (in rare case they may not), and they usually have the same Moon sign. But the rising sign can easily be different. I have experienced this first hand with two friends – identical twin sisters. The first sister I met dressed mostly in plain jeans, had a plain hairstyle and was much more into her books than her looks. I met the second sister sometime after knowing the first sister. The second sister was stunning, her hair was always done in a style. She mostly wore dresses and skirts and her entire demeanor was feminine.

I did both of their birth charts. And there it was, one sister had Gemini rising and the other sister had Cancer rising. Maybe you can guess who was who. The Cancer rising was the more feminine sister. It turns out that the difference between their birth times was about 15 minutes. And that was enough to have very different birth charts.

In the next example, we have two close friends with similar personalities but they are not genetically related. Both friends have Leo Sun signs. And the charts have a few powerful connections. The Mercury of one friend is on the exact degree of other friends' Sun in Leo. And the rising of one friend is on the same degree of the other friend's Venus in Gemini. This gives them a very broad set of similar traits and gives the friendship plenty of mental stimulation. They can talk for hours. The Venus and Rising being so close, gives them a natural familiarity. It makes them feel like they have known each for much longer than they actually have.

The last example is two lovers. The woman's Sun sign is Scorpio and the man's Sun sign is Leo. The woman's Venus is very close to the man's rising sign in Sagittarius. There are lots of initial sparks between them, both Scorpio and Leo are very passionate. But those Sun signs are classic conflicting signs. Still they have the positive power of Venus and Rising sign together, but it still might not be enough to overcome the conflict in their basic personalities.

PUTTING IT ALL TOGETHER 97

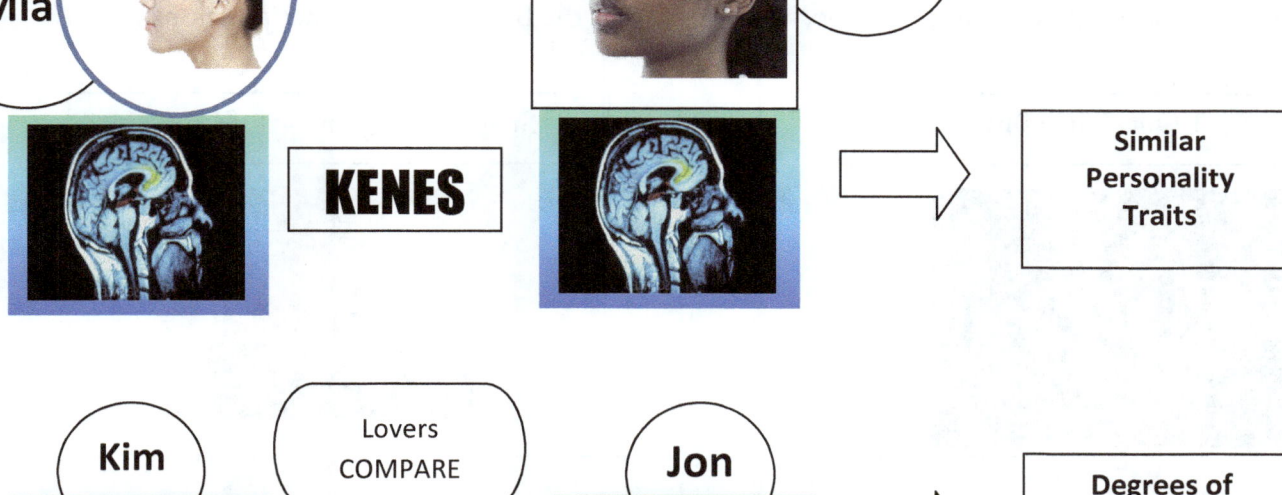

No so-called behavioral Science can do this.

*Genetic level, requires analysis of the genes at a molecular level

CURRENT MODEL SUMMARY

Each of our stories begins at birth, it is when we begin to interact with the outside world. We can see the entire process from its beginning stage to the ongoing life stage, and recognize the distinct differences of each. Since you are only born once, the second stage of ongoing influence does not get stored, it is a passing effect. The influences in the second stage are only present for as long the planets stay in place. This can be hours in the case of the moon, and many years in the case of Pluto.

One of the key parts in both stages is the sensory part. Somehow, we are able to receive these planetary forces. In the first stage they become stored. But during the second stage they are received, but only "compared" with what is already stored at birth. There has to be some kind of mechanism that acts to provide resonance for matching the planets' positions as they move through favorable and less favorable positions to the original position at birth. Like the children's toy of circles, squares and other shapes trying to match the "square peg to the circle hole", sometimes resonating favorably and other times not so favorable, comparing and combining the stored effect with the ongoing effect. In astrology, this is the basis of the personalized horoscope that superimposes the current planet positions on top of the birth chart's planet positions.

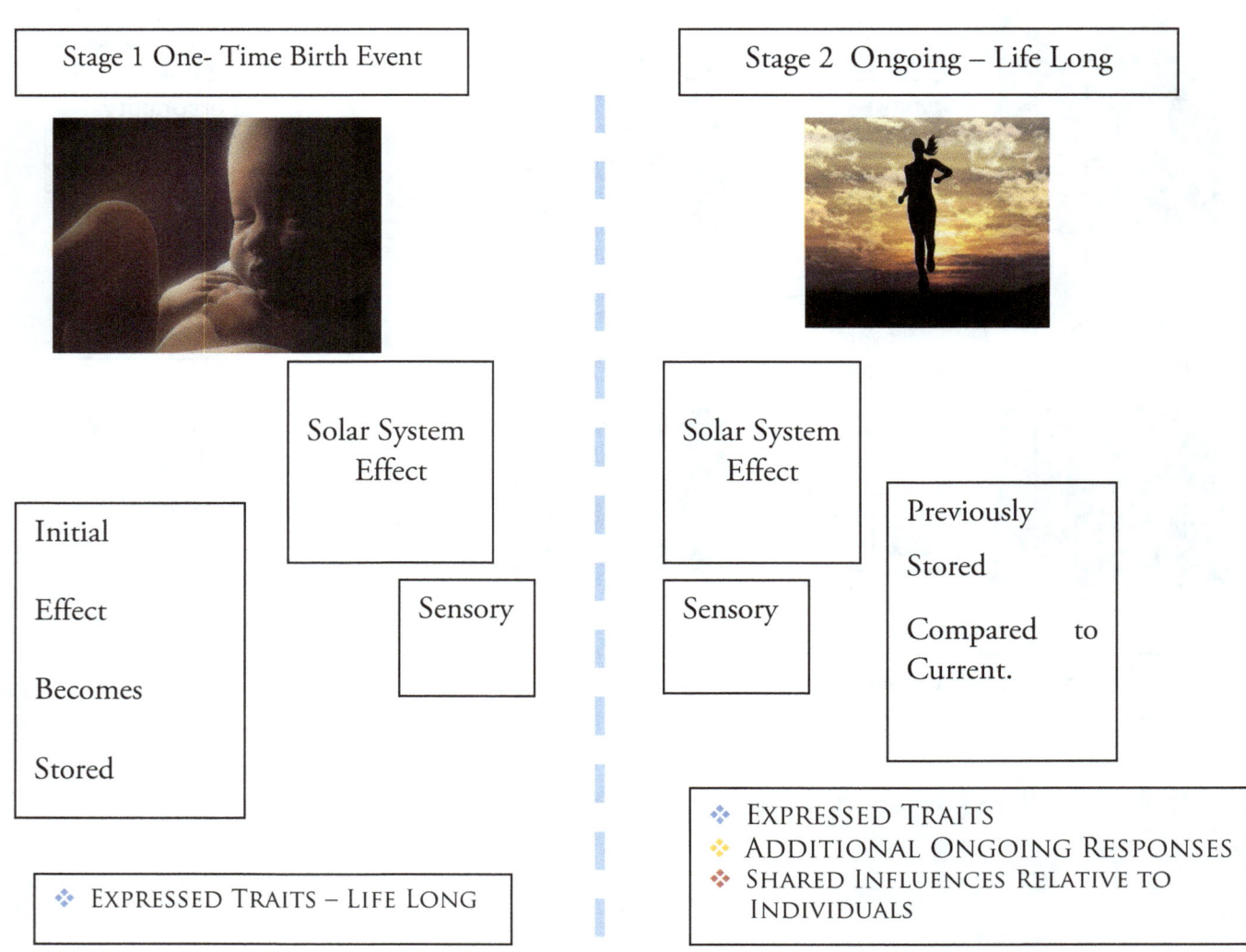

# Creating a Context

The phrase, "it's in the blood" has been used for hundreds if not thousands of years. Also the common idea and phrase "blood relative" has been corrected by genes. We now know that who we are closely related to is a matter of genes not blood type. Genes that are not just in our blood, but all over the body including hair and skin. But the whole "gene thing" has gone out of bounds trying to account for personality and intelligence. From our journey through the sciences, we have seen that our personalities are the result of cosmic influences and intelligence is part of that package. Personality appears to reside in our brains not genes. And while intelligence is also not genetic, environment plays a critical role in the development of intelligence. So which factor is more powerful, is it genes, environment, or personality traits from birth?

We start by creating two main groups to properly place each factor we know of. The first group of factors are fixed, because we cannot change them. There are three factors here, first are the genes we are born with that give our physical traits. Second, our personality traits at birth from the solar system, and third the early childhood environment. The second group of factors are dynamic and do change. The first two factors are amazing, they can modify certain parts of the fixed factors, these are epi-genetics, continual solar system movement, and last, but maybe most important, the third of the changing factors is behavior – your choices. This final factor is where all of the other factors can be overcome or fully realized, because this is where you take control and make all the decisions.

Our challenge is to figure out how each factor relates and interacts with the other to produce us and the lives we live. And what can we do to get the outcomes we want? That's really the whole point isn't it? But the interaction between our traits and the environment is tricky. If the environment is limiting, a personality trait may never develop to its full potential. The same is true with physical development, no matter your genetic endowment, if your environment is not stimulating, physical expression will be stunted. For example, maximum height is affected by diet. And yet it's a cop-out to say, "that a person is smart because they were born that way" - wrong. Development does not happen by default, no matter what traits you were born with. You must apply effort to develop yourself. The greatest people achieve results through focus and decisive action.

Let's start with how the fixed group works together. We have genes from our parents and that gives us our looks and our body. But we also have personality traits, and those are from the solar system, they are described in detail for each of us in birth charts. Personality traits range over all 10 areas of life, and even though we have them stored at birth, they emerge little by little as we age. For example, in the area of possessions and money you might have a trait for wastefulness, that trait probably would not be shown as a toddler. Genes will do the same, we won't see all of our potential right away. Our bodies will grow from infancy over time and our physical traits will flower into our full adult form.

CREATING A CONTEXT

On top of our genes and personality traits we are immediately immersed into early childhood culture and the surrounding environment. This will shape and develop us. Sometimes this can enhance other times it stunts potential. This is also where culture comes in. Culture can be described as "group habit", it has incredible reach. All of us have been influenced by it. For the most part, cultural norms are learned within the confines of close family relationships during early formative years of childhood, in two stages – from birth to 3 and 3 to 6 years old. What is learned and experienced during this time becomes deeply rooted and plays out through the rest of life.

It's very common to mistake the powerful effects of culture for genetics. For example, "it must be in those people's genes to act that way". Behaviors have virtually nothing to do with genes. It is culture that is behind the actual transfer of certain group behaviors, and ideas from generation to generation. We simply learn the habits of speech, diet, household norms and a range of other actions directly from our families. Culture continuity can be extremely strong considering these are early learned behaviors, but they can be challenged and altered later in life. We see within the same culture many individual expressions, varied interests, likes and dislikes, and other tendencies. Even within strict households one child may show interest in music while another sibling has a talent and interest in sports, yet both children are raised as protestant in the suburbs.

However, the stage of acceptance during adolescence can drastically alter some of this cultural indoctrination and cause the adolescent to challenge the validity and value of previous cultural norms. In multicultural environments some biases taught in the home may prove to be invalid or the ideals of other cultures may prove to be beneficial or compelling in some way. For example, general interests like food, dance, and musical tastes mostly come from culture along with a whole range of social viewpoints and behaviors, none of which are genetic, yet they can be altered by cultural exchange. Let's be clear, culture does not directly shape personality. Culture shapes mentality, opinions and viewpoints. Personality is much more "personal", while mentality is a broad view of the world and are behaviors practiced by many.

Our experiences can vary greatly within the same environment, even within the same home. For a number of reasons, siblings often have very different experiences in the same household environment. Age is an obvious factor as to why parents may treat children differently. Whatever the case, experiences can be very different in the same environment.

Many behavioral traits are complex, often only being expressed as responses to certain environments, situations or actions and the traits of others. But other traits such as musical taste can have multiple sources. For example, you may be raised and exposed to only one genre of music, say country music. That is determined by environment and culture but your built-in personality may determine specifically which songs you enjoy the most. Just because someone's birth chart shows a tendency for a certain type of behavior doesn't mean that person has no control over this tendency, there is always choice and responsibility for your actions.

If we start off with physical traits and personality traits that we cannot change, and are born into an early environment we did not create then it would seem that genes, birth personality traits, and early environment must be the strongest factors, right? Not necessarily, nature has given us plenty of opportunity to choose. However, if genes have no defects and a child is generally healthy, the early environment may present extreme disadvantages or extraordinary advantages and may be the single strongest factor in an individual's life. Obvious examples would be the prison of abject poverty or the monarch-like privilege of one-percenter wealth.

The constant movement of the planets in our solar system produces an influence similar to how the birth moment gave us our personality traits. The difference is these constant movements are focusers and amplifiers that can be helpful or difficult, positive or challenging depending on where you are in life and what your exact circumstance is. If you are alone in Siberia for a month doing research and a horoscope forecast says that this is a time for a new romantic interest, that seems highly unlikely to happen in isolation. But at the same time if the chart indicates a major breakthrough then maybe you are in the right place for it to happen. Even genes are not a closed set of instructions they are also open to environmental and behavioral feedback. Choices such as diet and exercise can change genes. This area of science is called **epi-genetics**, "epi" meaning on top of.

Outside of the extreme cases, it is personal behavior and action that is the most powerful factor. After behavior, great genes and or great personality traits when used properly can lead to many advantages. Across humanity the appearance of great traits is more or less random. Some individuals are simply more able than others to easily grasp advanced science and technology, demonstrate extraordinary talents in the arts, show top level athleticism, or be visionary in leadership or industry. Even though these traits may have been there since birth, they are still secondary to personal action, will, discipline and hard work.

In conclusion, on a basic level, life can be broken down to decisions and traits interacting with the environment. Life then becomes a matter of how we decide to use our genetic traits and personality traits to get the outcomes we want. Therefore we can increase our chances of success by becoming aware of how our decisions and traits are going to interact with the world around us.

TEMPLATE OF LIFE

Now we can assemble all know factors of life into one template. Looking at all of the life factors shows us that there are many ways where constructive and positive intervention are possible even if negative circumstances were present at earlier stages. Some factors of influence may affect given areas of life more than others, however *no single factor alone determines life, they all work together.*

FOUNDATIONAL FACTORS

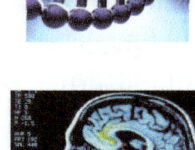

GENETICS
The legacy of our physical make-up, and appearance passed down to us by our parents. Genes set parameters for your physical traits such as eye and hair color.

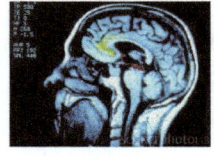

KENEMONICS – Solar System Position at Birth
A basic set of traits comprising your core personality. Each person is given a set of traits and tendencies. Knowing these up front simply serves as a guide to maximize one's positive attributes while identifying areas of fault to manage. It is much easier to develop known talents and attributes. And "success" in life is often a matter of managing one's faults. Yet these attributes do not determine your life, rather by acknowledging them you can harness them and shape outcomes of your choosing.

DYNAMIC & CHANGING

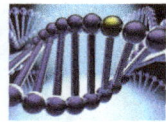

EPI-GENETICS
Environmental and behavioral factors that modify genes and or gene expression. So even our individual genetic blueprints are not completely "set in stone".

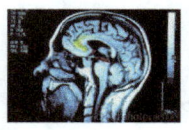

KENEMONICS – Solar System Movement
The same planetary forces present at birth continue their cyclical motion and also continue to play a role throughout life - a continual environmental influence.

EXPERIENCE SET
Includes the culture we are born into and all other life circumstances that we do not control combined with the choices we make. The Parents you were born to and the overall rearing environment. The influence of culture is extremely powerful, we are all products of culture to varying degrees. Circumstances that are always changing in every context not only socially but all other areas of life.

BEHAVIOR
Life does not just happen to us, we are presented with decisions. We make decisions to act, research certain topics, pursue formal education, self-education, diet and exercise, travel, our treatment of others, cultivation of relationships, and cultivation of our character. Behavior is not just output. Actions often cause direct reactions, these reactions in turn come back to affect you as part your life experience.

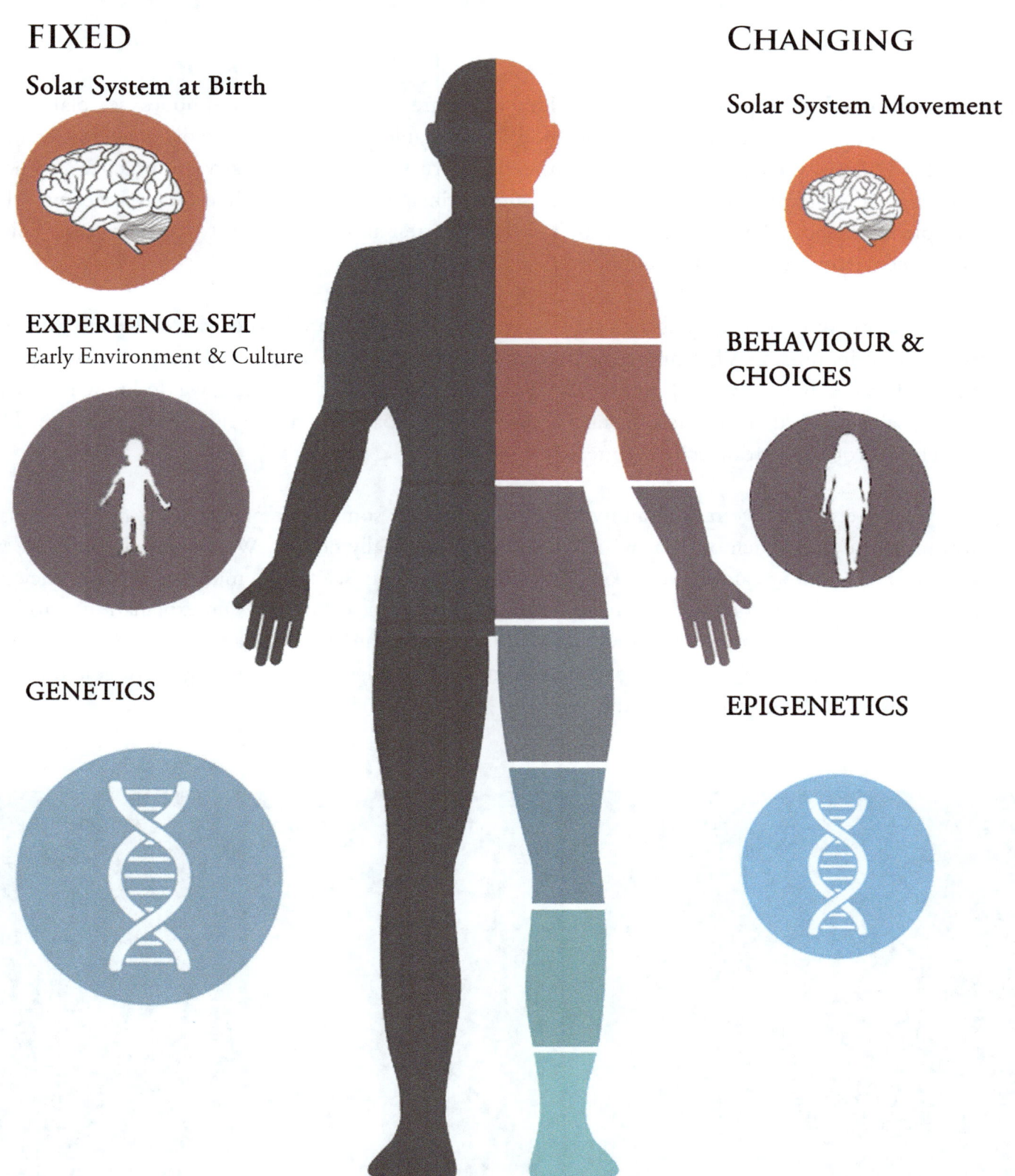

PART THREE

FRONTIERS OF MODERN SCIENCE

We blindly accept the mainstream idea of progress. We have smart phones, jet planes and microwave popcorn. But we do not have global sustainable energy, a sustainable economy or even clean food, clean water or clean air to breathe. Modern science and industry are practiced without concern for long-term effects. So we practice some principles of science while totally ignoring other principles of science. Then we crown this era as the smartest and best humanity has ever been. This is a monumental and tragic contradiction.

Mainstream science is largely a tool of industry driven by profit and war, so we are generally not aware of its misses. What progress, inventions and discoveries would we have if our system was not based so heavily on war and profit? We don't really know. But one thing we do know is that we have missed the fact that the ancient practice of astrology is based on a very real physical science. This has to be one of mainstream science's greatest misses.

What else have we missed and why? It's time to ask some basic questions. It may come as a surprise to you that much of what we think we know, we really do not. We assume that because we have cable TV and streaming video on our phones that "our scientists" must know what they are talking about, right? Wrong, let's take magnetism. Mainstream science gives no explanation as to how two magnets move each other. But more on that later. Even though, there are legitimate new areas of science, like Nanotech. We are going to expose artificial intelligence, genetic engineering, and several other fields for making fraudulent claims.

SECTION 1
MODERN SCIENCE

Why don't we consider our personalities affected by the solar system? Why don't we see the connections where the ancients did? The main reason is the erasing of science's lineage. In the 1700s a movement started, an avalanche of lies, to bury and erase the traces of the African, Arabic and Asian roots of science. Figures like Carl Von Linnaeus, so-called father of anthropology, in his mockery of science, that included no science at all, provided his basis for racial classification. The idea of European racial superiority became the academic norm. This idea was injected into all areas of science, history and religion. Others like John Locke and Adam Smith would further a material view of the world, restarting a system started by Plato. In the 1700s thought leaders of the west omitted the scientific progress of 711 to1400s, from their history books. This denies credit to the historical achievements of the Moors, Asians, and Arabic societies.

The movement of omission even included other Europeans. After 1727 the British Royal Society blocked the publishing of Isaac Newton's works on Alchemy. They proceeded to totally remove alchemy, spirituality, and philosophy from science, making science totally mechanical and material. By doing this vital values and ideas were lost and the genius of humanity has been narrowly directed to develop machines ever since. There has been a constant omission of any figure or achievement that is not western or European. The mainstream celebrates the Greeks as pioneers, but the Greeks gave direct credit to Ethiopia and Egypt for being the source of great ancient knowledge. To quote Aristotle himself, "and thus Egypt is the cradle of the mathematical arts".

As a result, we are left with a false narrative. The contributions of non-westerners are edited out of the human story. The true identity of those at the source of civilization go nameless. To this day Africans are not associated with their inventions of steel, medieval and ancient optics. Without Lewis Latimer we have no light bulb filament. Garret Morgan gave us the traffic light, and Thomas Mensah's 7 fiber optics patents - no mentions. The western thought leaders of the 1700s did a much more thorough job of destroying traces of non-western scientific origins than even the Romans burning of the Library of Alexandria. To this day, many think the world was waiting in darkness until Europeans came to everyone's shores and invented everything we see. In truth African and Asiatic ships had already touched every continent, in some cases thousands of years before 1492.

Back to the point, what was lost in this process was the ancient inclusion of spirituality. I believe loosing this sense of connectedness between all things is one of the greatest tragedies in the last several thousand years. It has directly led to the destruction of the life support system on this planet. Western Science, with the participation of a diverse human cast, has gone on to produce computer and space technology and the internet, these are incredible contributions to humanity. However, just image the world we would have now without the false concept of European racial superiority and without the wholesale destruction of the environment.

Paradigm Shift

We are coming to a point where the current system's practices of the old industrial revolution are visibly failing. Science and knowledge have out-paced those old industrial based practices. The old ways are having consequences we cannot ignore. Climate change is in the headlines everywhere. But what is not in the headlines is the underlying principles or lack of principles that got us here.

The scientific method, for all of its prowess in determining the truth of a claim, is feeble when it comes to principles of consideration when practicing science in the real world. It is a method missing the essential principle – connectedness. The deeper we look into the problem we see that "science" isn't really the problem at all. It is the modern version and use of science. The source of the problem is our disconnected westernized cultural ideals. You can see it everywhere, "destroy something like a rainforest over there to fuel an industry over here." This destruction is assumed to be part of human nature, this is flatly wrong. This destruction of nature is only widely imitated around the world because it is falsely seen as the route to prosperity.

The solution is to replace the current system with a system of connectedness – the consideration of all things. The idea may seem crazy - to merge science with other ways of dealing with life, like ecology, philosophy, and even spirituality – all combined together. But if our "great minds" and "advanced science" can miss that our current idea of progress is actually a side by side march of destruction for each step of advancement. It's time to reevaluate what we are doing and make some serious changes.

What does any of this have to do with astrology? The question is, how did we miss the valid science behind astrology in the first place? The issue is much deeper, but the answer is simple – we lack a sense of connectedness. Astrology is shunned by mainstream science. Yet astrology represents a connectedness at a fundamental and cosmic level. So it is fitting that astrology is shunned in such a disconnected society. I believe we are on the cusp of a major discovery, connecting human personality to the planets. At the current rate of discovery in the field of neuroscience, there is no doubt mainstream science will be forced to admit, "that yes, we are influenced by the planets." This will cause a huge shift in the way that people view themselves, but before that can happen definitive evidence has to be found. So let us continue our search.

SECTION 2

REMAINING MYSTERIES

Most of us take for granted the basic science taught to us from elementary to high school. We do not question it. Amazingly, something as basic a magnetism, at a fundamental level still remains unexplained. Yes, the simple magnet. Maybe this is the reason for the lack of ability to connect gravity with the other basic forces? This is a prime example of a phenomenon shown, but not fully explained. Many great discoveries start out with something that everyone can clearly see but cannot explain. A great example is inherited physical traits. People thought it was blood that passed on these traits and the guessing continued. It was a mystery - how and why we look like our parents. Blood was wrong. Eventually our techniques of investigation increased to the point where we actually discovered the DNA molecule, and the field of genetics was born.

Next we start the search to explain memory in the brain. The mystery of how our brains store information is key to explaining how personality traits come from the solar system. And with increasing abilities to scan the brain's structure at the molecular level, one day a configuration will be found. And when brains are compared just like DNA is compared, it will reveal a configuration that corresponds to birth dates for people with similar configurations.

I'm going to come right out and state it, mainstream science is wrong about magnetism and gravity. It's one thing to present truths about astrology and challenge academic scientists on facts about astrology, with which they are not familiar. But it is completely unexpected to attack academic scientists where they are most strong – physics itself. Mainstream science has been asking all of the wrong questions and looking down the wrong paths. The invisible forces of the electric field, gravity and magnetism, are some of the most mysterious areas of science today. Wind was once an invisible force, and you can include sound with that, until we discovered that there are tiny particles of air.

What is the nature of the remaining unseen mysteries, current science is infant-like in explaining these. Mostly they just don't bother to mention the question at all. Either they don't know or they are intentionally not telling the truth. Which is it? In any case, this giant missing explanation should cast a serious shadow of doubt on mainstream science's self-proclaimed mastery of the physical world.

Magnetic Deception

We have all used magnets, played with them as children, and hung things on our refrigerators with them. What exactly is behind magnetism's invisible pull and push? What is moving the magnets? Most of us know that like sides repel and the opposites attract, but that's about it. Let's turn the table on the science that laughs at astrology and demands a full explanation of how a planet can affect a human. Yet they cannot tell us how a magnet moves another magnet.

It is obvious, that if an electric fan has blown papers from a desk to the floor it is not the fan that moved or pushed the papers off the desk. The fan is making air move that in turn pushes the object in front of the fan. The same with a vacuum in the opposite direction.

How does the same object that causes a pull also cause a push? For something that has been so assumed as basic magnetism, no physics textbook on any level of education even tries to touch this subject. Something is not right here. The entire subject of magnetism is presented in a fragmented and incomplete way deliberately leaving out the fundamental questions, which should be obvious. "What is going on in between 2 magnets to make them move?"

That question is never truly explored. The most we get is a lame answer about lines of force. So why? Is it either because the people behind classroom science don't know and don't want to admit it, or some may know but are deliberately trying to keep the answer buried?

The question is not irrelevant or trivial. So don't be scared off by the big scientific terms, these are everyday things. Understanding the invisible forces of magnetic field (any simple magnet), electric field (static cling on clothes), electromagnetism (light, radio), and gravity is the key to understanding how the planets affects us – and to fully explain Astrology. If you don't understand what the big deal is. Don't worry the "ah ha" moment is coming.

The textbooks use terms and concepts like, lines of forces, eddy currents, magnetic flux, magnetic field strength measured in gausses – none of these terms remotely deal with or explain the most basic question. What is happening around a magnet, what is the magnetic field made of? The explanation of the 3 cases on the next page are always completely and possibly intentionally left out.

The text book explanation is that the lines of force are directional, coming out of one end (north) and going in the other end (south). This still is not an explanation.

Mainstream science explains gravity by saying large objects bend space (space-time). But is a magnet bending space? No, magnets are clearly not, because two ordinary magnets, with a pencil between them will not affect the pencil but will pull and push each other. So what is happening between the magnets? They have no good answer. Modern scientists are not used to being called out on flimsy explanations, but are quick to brand astrology as pseudoscience. What we have is hypocrisy, a double standard. Modern science asks, "Explain how the planets can affect us?" We give them the birth chart. When it comes to magnets we ask, "Explain how magnets attract and repel across empty space?" We get no answer.

We can't wait for them to figure it out. So without solving the mystery here and now, there are a few things to consider. Very close observations of magnetic field suggests there is motion, interlocking, the field has an internal shape, and a scale at least as small as a molecule of the magnetic material. Let's go over the three basic cases of magnetism.

CASE 1
North & South Attract

Lines of Force

CASE 2
North & North Repel

Must be different?

CASE 3
South & South Repel

In Case 1: Opposites attract, clearly north has something going on different than south – the result is they come together.

In Case 2: Two norths have the same thing happening on both ends. The result is they repel.

In Case 3 : Two souths have the same thing happening on both ends. The result is they repel.

PROBLEM: Case 2 and 3 are not the same ends and must have different interactions, but they have the same result – they repel!

Again, gravity is supposed to bend space itself. Even with gravity no object attracts or repels by itself, there is always some medium acting in-between objects. So is there a medium acting in-between magnets? There certainly seems to be, and it is an issue we will explore later.

Gravity

The mainstream science approach to gravity is either deliberately false or another huge cop-out. The Earth pushing down space like a bowling ball on a mattress, gives the wrong visual impression. A more correct way to picture gravity would be waves of giant bubbles starting from far out in space collapsing into the Earth's core. A naturally occurring spherical gravitational body doesn't fold space in a linear fashion not even close. And bending of space-time doesn't account for the obvious lack of pulling everything in space into one giant ball.

Also, it is fairly clear there is an effect that occurs around the equator of spinning celestial bodies and a balance resulting in the separation of planets, moons, solar systems even galaxies. Bending space-time may be a byproduct, component or co-factor, but it is being presented as the grand explanation. Below is the false 2D mattress grid visual (over simplification) of space-time pushed by today's classroom physics teachers.

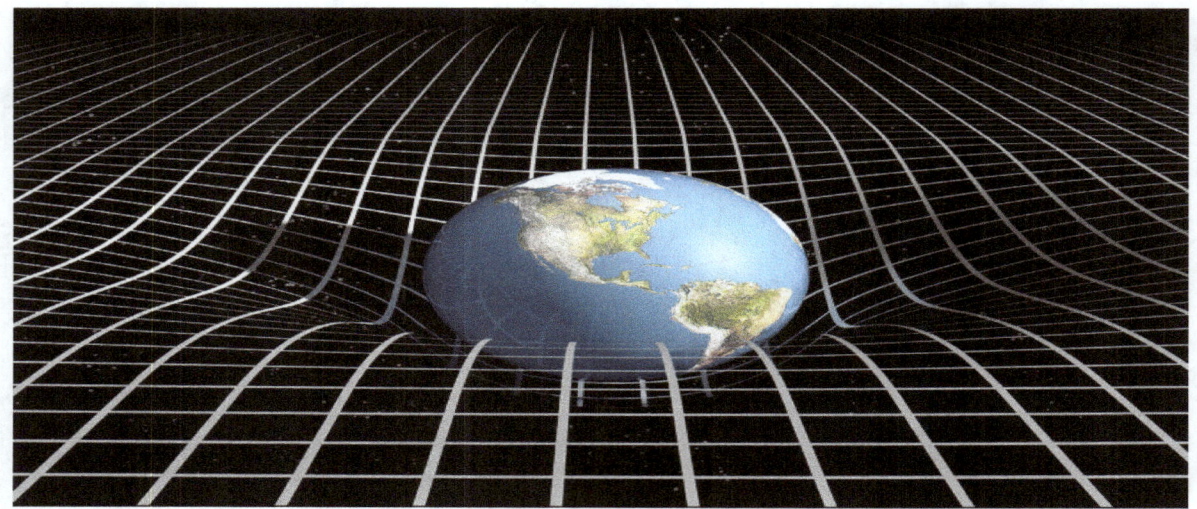

INCORRECT / INCOMPLETE

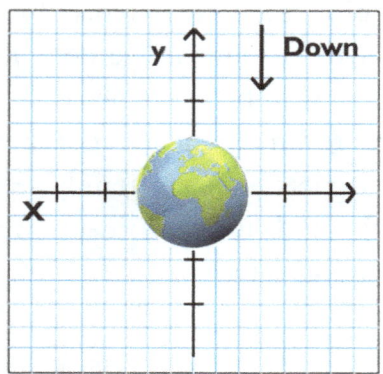

MORE CORRECT

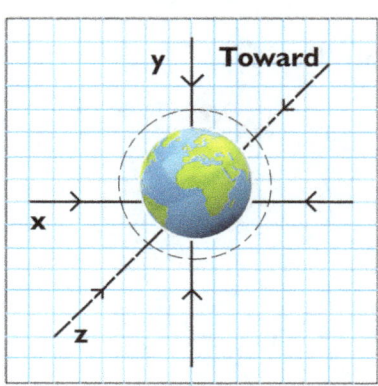

We are also told gravity is only a force. Maybe the fact that it is more than that is a closely held secret? Picturing gravity as shown below, makes a point of a spinning planet's most interesting gravity feature - the gravity disc. Highlighting this shows that the earth is divided into two gravity areas showing gravity varying in direct from pole to pole flipping at the equator. This is key for understanding axis tilt alignments, the alignments that set up the zodiac signs and the seasons.

The north half of Earth is pulling in a different direction than the southern half. Astrology's birth chart hints at having knowledge of this by requiring the exact place of birth, requiring this information because it makes a difference. So yes, Earth's gravity does pull towards its core, but there is literally a twist in the pull and a flip in direction on either side of the equator.

The simplest way to think of gravity is a wave of bubbles collapsing toward a planet's core. But if you put everything that gravity is doing in one picture its going to look crazy and overcrowded. But we can put the most important things in the picture. There is too much particle physics going on, we need field studies. What happens when one gravity field goes through another gravity field? What happens when gravity pierces an atom? What happens when a gravity field passes through a magnetic field or an electric field? Let's start combining fields and EM and see what happens. All of these ideas are is completely ignored by mainstream physics.

GRAVITY

Electric field

PUSH & PULL

Magnetism

PUSH & PULL

Gravity

TOWARD

The forces of magnetism and electricity once only observed as separate. Both have push and pull, now used to do much more! Pictured below electricity and magnetism in a current inside a wire.

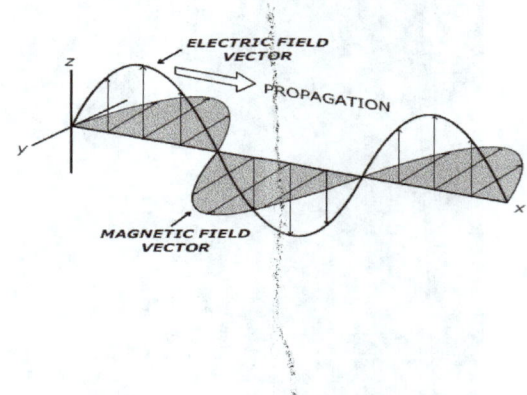

Can we link gravity to electro-magnetism beginning with basic observations? If we think of gravity as transmitting and receiving, we can say all things receive gravity but only relatively massive things transmit gravity in any meaningful way. Also, a planet like Mars does not need to affect you directly. Mars only has to affect the Earth to affect you. The Earth's gravity field will fluctuate, and that will affect you. For example if you were on a raft in a small lake or pond and a large enough object is dropped near the raft the object doesn't have to contact you to affect you. The waves created by the object will.

If gravity results from a large collection of matter then gravity is in all matter but just has to reach a threshold, an extreme mass, or amount of atoms before it is powerful enough to interact on a planetary level. There is also the threshold of number of brain cells or complexity of connections and number of connections before the brain is able to sense and connect with the world around it near the human level.

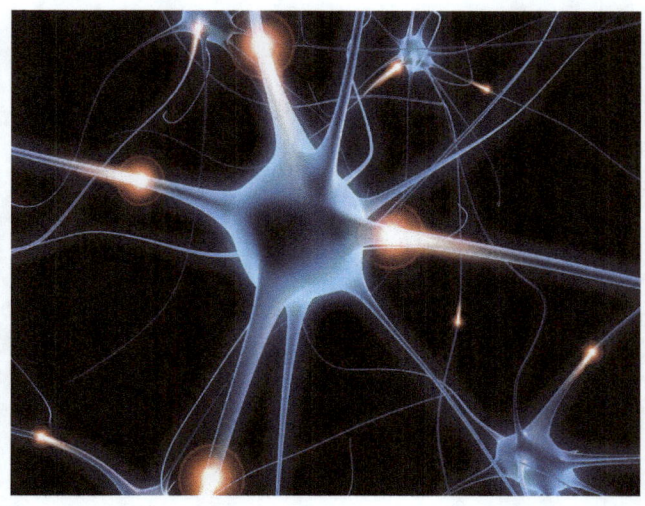

Worm: hundreds of nerve cells

Ant: 250,000 nerve cells

Cat: 760,00,00 nervous systems cells

Human: 86 billion brain cells – high intellect potential, individual personality

1 Human: 7×10^{27} atoms

Earth: 10^{50} atoms (1 billion times more atoms than all humans combined, times 100,000)

Sun: 10^{60} atoms (approx. >1 billion Earths)

On one hand many prominent scientist boast about the fact that our bodies' atoms were born from stars. Yet at the same time dismiss and never truly explore meaningful ways we connect to the sun, planets and the forces that come from them. The obvious position seems to be that we would have a more meaningful connection to the actual source of our very life substances.

Attraction

Attraction is everywhere and on every level of life, not just within the texts of science books to be measured in some mechanical way, but in our everyday personal lives as well. Everything is in motion, and when you look closely its usually because of attraction.

First, let's examine what mainstream science tells us about attraction and what is missing from that version. We are told that gravity is also a force of attraction. But unlike magnets, gravity tends to create configurations of motion like the solar system, and does not always just "connect" the parts like the attraction of two separate magnets north & south placed close to each other.

The atom also has attraction that results in configuration. The negatively electron is attracted to the positive charge this allows for all of our gadgets, computers and vast arrays of technologies to work. Attraction has its opposite - repulsion. In magnetism both attraction and repulsion are easily seen.

Also similar to gravity, magnetism only appears with a "collection of matter". Magnetism seems to contain components of charge, gravity may contain components of magnetism, on a field level. All of the basic forces and fields must have some components of the other forces inside of them otherwise there would be no reaction between them.

Gravity is a system of "tethering", despite the idea that all things pull on all other things, the basic concept of gravity fails to reveal that all objects have a primary link or tether to the nearest most massive object. The easiest case is the Earth, Sun Moon relationship. It's obvious that the earth's orbit is centered around the sun, which makes the sun the earth's primary link.

The moon is basically the same distance from the sun as the earth, but the moon does not have a separate orbit around the sun, it has an orbit around Earth. So the Moon's primary tether is to the Earth. So even though, all of these cosmic bodies "pull" on each other the relationship has order. So our primary tether and everything in earth orbit is to the Earth. The Earth's primary tether is to the Sun. The Sun's (and the solar system) primary tether is likely the center of the galaxy where there is a massive black hole.

Important Observations

❖   All of the fields have a motion or movement component. Current or moving charge can produce magnetic fields. Moving magnets can produce current which produces charge. Those are well known, but gravity can also have spin, this is extremely important. There does not seem to be any similarities between gravity and magnetism until gravity is spinning.

❖   Are spinning bodies showing us a form of gravity repulsion to each other when the scale is close in size? Galaxies, solar systems, stars, and planets all spin. What sets the spinning in motion in the first place, why is it there? Is spin present on the smaller scale forces like magnet field and electrical field, but we just can't see it? We know that electrons have spin, so it's not such a leap to think the fields could have a spiral form at some extremely small level.

❖   Gravity has *poles* because of the gravity disc. The sides don't differ in pull, but they do differ in direction. The gravity of spinning bodies also allows for a type of balance. Moons are caught in the spin of planets, planets are caught in the spin of suns. The result is, the solar system stays separate.

❖   There is also the gravity paradox - attraction but no full attraction. They explain this away as saying the orbits are in decay, yet they are still being pulled just at a safe distance, that has to mean balance. The act of a gravity field being able to interact, balance even repel cannot be ruled out.

❖   In atoms the opposite charges do not join like in magnets. Negatively charged electrons are all together in a cloud of orbitals, separated from the positively charged protons in the nucleus.

❖   Possible magnetic field shape: Opposite coil shaped force at each end. The north and south poles would fit together like a screw in a thread. But north and north, or two souths, pushing them together just compresses both coils that won't interlock, so they push each other away when you bring them close.

❖   With magnets, it is impossible to separate the negative charge from the positive charge, in one magnet.

❖   Almost countless divisions of both a holographic plate and a magnet. Combing of magnets, maybe the same for gravity and some holograms.

SECTION 3
HUMAN BEHAVIOR

In the world of science, human behavior remains a puzzling mystery. The current practices and clinical observations of human behavior are incomplete. They lack the full template of life and do not factor in that personality traits are given by the solar system. But still, mainstream behavioral science can be useful, even with an incomplete model. Behavioral science could be greatly enhanced with the knowledge of born-in personality traits. Evidently, this has already happened. Just like other areas of science, behavioral science ignores the historical roots of their own field, refusing to credit the role astrology played for the founders and pioneers. Carl Jung, an early pioneer of the field used his patient's birth charts to help analyze his patient's conditions.

Why was Carl Jung compelled to use the birth charts of his patient's to aide his work? It is because the birth chart is all about human behavior. The birth chart was able to give Jung insight into the basic make-up of a person. Without guess work, Jung could then perform direct observations more accurately. If the birth chart had nothing to offer, he would not have used it. He confirmed the birth charts usefulness. Remember, there would be no astrology if human behavior did not confirm it. Human behavior is the biggest proof of astrology's validity, and it will probably always be.

The behavioral scientists

One of the problems with behavioral science alone is the lack of explanation for the difference in the base personality, child rearing and birth order are some of the default reasons. But what about when the environment is virtually identical, of course the experiences always differ, even with twins. But with the current textbook explanation there is no accounting for a more base level of personality that comes before in-home social relations. Carl Jung, combined astrology with psycho-analysis, using the birth charts of his patients. This is an inconvenient fact carefully silenced by today's so-called behavioral scientists.

Astrology's birth chart often describes young siblings' unique tendencies within the same family, but this doesn't get included among the accepted factors in mainstream practice. To some extent it doesn't matter, because through proper analysis they can determine personality in each specific case. Away from their practice some may acknowledge the validity of the birth chart. A wise choice given the backlash that can result from openly straying from orthodox practice.

Fortunately, proof seems to be on the horizon, a discovery of actual individualized brain structures, however small, could change this view. But even then, caution should be taken not to overreach as genetics has. As we have seen with genetics trends alone can lead to acceptance of ideas more than actual facts. We must take caution to avoid a predetermined type of outlook for individuals. A properly balanced outlook must be fostered between all the factors of life. After all what is the overall purpose? Is it not to enhance well-being and self-knowledge to empower the individual over undesirable patterns of destructive and self-destructive behaviors?

One thing is true, the factors that go into the formation of individual personalities are complex. However, we have to consider the complete picture. And only Kenemonics as with astrology is able to provide an actual framework and basis for compatibility that eliminates much of the subjective murkiness of analyzing individual personality traits. All of these areas where astrology has thrived and now Kenemonics thrives.

Searching in Vain for behavioral Genes

The DNA molecule was discovered in 1869, and in 1957 its structure and function as the carrier for hereditary traits was demonstrated. In 1984 Human Genome research planning began. In 1990 the Human Genome Project officially started. In 2003 the complete genome was sequenced, a total of 6 billion characters, 4 letters, and 20,500 genes. Zero of the genes have been identified as absolutely behavioral. Many have been proven to be for hair color, eye color and a huge range physical traits.

I was working among the companies involved during the project, it was exciting. Optimism was high, a lot of money was poured in. The genome was achieved. Then even more money was poured in, and a lot of it went down the drain along with the hopes of many large companies that sought to cash in on the human genome.

This long and intense period of research and discovery has led to thousands of genes being identified. The idea that human intelligence and behavioral traits also owe their origins to these genes also entered the discussion. Overnight, genes became the de-facto "ultimate source of all traits". Like a scene from a movie, the entire establishment bought it, as the usual suspects of science refused to accept reality "Personality is here, it's in genes. There's no place else, it's got to be here!" But this idea is just not true. Basic observation can tell you that a child "looks" like its parents, so the idea that children receive physical traits from parents is an observable fact, now 100% proven by genes. However, when it comes to behavioral traits and different types of intelligence there is no obvious concrete observation. The idea that genes endow the individual with a personality is just an idea. Unfortunately, it is an idea that is a search in vain. As search similar to searching in the middle of the Pacific ocean for fresh water.

Undoubtedly, genes are the blueprint for the physical body, but not the code for human personality or intelligence. Many companies that have helped produce the human genome have found the process of creating gene based drugs elusive, and as a result most of those companies are gone. To me this is further proof of overreach and inflated hopes of the field of Genetics as the end all be all for human traits. While having the genome of a species is extremely helpful in creating a framework for understanding the processes of cellular development. And fields like Bioinformatics are beneficial to enhancing medical treatments. By no means, should genetic research be reduced, to the contrary, our efforts and participation should increase in genetics and its sibling fields. However, the direction and hopes need clarity and a systemic re-check.

So bring the focus back on track, the seat of behavior is the brain. Of course, if you change the quality or function of the brain you will have changes, but these are not behaviors in the true sense. For instance, if there is a genetically linked deficiency in serotonin then there will definitely be an effect on mood, but that is not nearly what is meant by built-in personality traits. Also, if there is some sort of genetic "defect" that cause the brain not to fully develop this to may have real effects on behavior and "intellect" or cognitive functions. But again, these are not part of forming one's natural personality they are more basic issues along the lines of physical impairments not formation of personality.

Elusive Intelligence

What is intelligence and where does it come from? All forms of intelligence, and there are several, have to do with sensing, perceiving, and storing information. And then processing information to act and create objects that improve the current condition in one's environment. So all intellect is linked to environment. The development of intelligence is either enhanced or limited by environment. Also, intelligence is randomly spread through all populations, just as we see it in everyday life no matter the socio-economic background.

Furthermore, intelligence is also a political tool, used to change perception about groups, legitimize some and demonize others – both are invalid. On a socio-economic level the issue of intelligence centers around power and issues of control and superiority. This is why we see so much desire on the part of the rich to engineer genes to create super smart children. But we are all children of the solar system which is the source of base intellect. So there will be no genetically engineered intellectually superior class exclusive to the rich.

The two main factors of intelligence are Kenemonics and environment. Provided there are no defects or injuries to the brain. It is very difficult to overcome poor schooling and adverse home and community factors. This is a motivation killer not an intellect killer. Even a well-endowed intellect must study, focus and sharpen their abilities. There are examples of early childhood top performers in impoverished areas that leave school but often end up achieving later in life, even if it meant going to prison where the books and education are in some cases far better.

Types of intelligence according to Astrology's birth chart.

- ❖ Creative Imagination
- ❖ Analytical
- ❖ Verbal
- ❖ Invention innovation
- ❖ Artistic expression

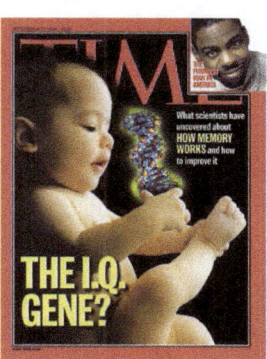

Another sad joke. Looking for an I.Q. Gene is going in the wrong direction.

In summary, human intelligence is a combination of two factors. The first is our base intellect, which come with traits given to us at birth by the solar system, and second the developmental environment, which depends heavily on early home and school education. Something else to consider, is that intellect comes in many forms. We should learn to place value on all forms of intelligence and gain from the combination of creative and analytical approaches, the merging of insight and wisdom with technological innovation. Clearly, the idea of racial intellectual superiority is false. You are not better because you are white, or black or Asian. Intellect is a cosmic event, randomly endowed across humanity.

Epigenetics Overlooked

For a long time genetics was viewed as a static thing. The genes you get are set in stone and that's it for the rest of your life. But careful observation was showing that certain behaviors might be altering the genes that supposedly dictate so many physical things. Genes are the physical blueprint, it's true, but even more so an interactive blueprint. Genes are also a code for adapting to life on Earth. This is well-known but often overlooked. So much of the current focus is on exploiting the genome to make commercial drug treatments. Genetic change is the area scientists need to focus on, not on behavior, but how certain actions and environments can alter gene expression.

Fast forward to the current day, the status quo of static and unchangeable genetic destiny and influence has changed greatly. The history of this resistance to change is telling. It's a lot easier for people to have an all or nothing mindset. And it takes enormous effort to accept the subtle changes that were being observed. There is a complex interrelationship between genes, environment and behavior.

The effect of some so-called "bad genes", can be changed with changes in the immediate environment. Behaviors like regular exercise and improved diet do not only change the visible body but the deeper cellular levels as well. Even thoughts, mediation and emotion self-control, can have large effects deep with the cellular processes. Studies are showing that consuming more caffeine can change how certain genes are expressed. There is still a question, does this only change you during your lifetime or can you pass these changes on? All of this may seem intuitive now, but it took a considerable amount of time and effort to change this general way of thinking. They teach epigenetics now in schools, but the common person who is aware of genes knows very little if nothing about epigenetics, people tend to think genes are forever.

Neuroscience & Brain Research

Currently there are efforts to map and model the entire human brain and represent each synapse, each connection and every possibility. The problem with this is that we each have individual configurations and the model that will be generated will be generic, and without individual meaning.

However, there are a number of possible benefits that may come with increased understanding of our brains. We may be able to improve our ability heal certain brain damage. And we discover direct ways to treat, maybe even cure certain mental illnesses. But the challenge to map all the possible neuron connections, may be impossible. If a sufficient general mapping of the brain is achieved we may still learn enormous amounts of new facts about our brains and what we are capable of.

Possibly the greatest prize of all are the kenes of Kenemonics. each personality trait, the true source of individual personality will not be known to the mainstream model. Presently, mainstream science is completely unaware that the solar system gives us our personality traits. As a result, the main goal of research is to copy our brains.

Fortunately, without a knowledge of kenes any copy would be incomplete. And our brains are not just processors of sensory information, like sight and sound that can be represented as data. The brain not only uses electrical signals and synapses, but also a complex set of biochemical, neurotransmitters and specialized receptors. The combinations and complexities far exceed simple zeros and ones inside of our best computers. Our brains have varied states of mind, emotion and consciousness that alter perception and processing of information. No matter how many supercomputers are used to simulate brain functions, there may be aspects of the nervous system that go far beyond what any semiconductor and microelectronic device could ever produce. You cannot simulate what you cannot fully understand.

Neuroscience & Brain Research

In all of humanity there are seven and a half billion brains with nearly identical physical structure. Yet all of these brains are filled with different memories and personalities. The difference is beyond structure. The difference has to be the energy and fields inside the brain.

Given that revelation, we now have a lot to look for. We can shift from only looking for physical substances, cell structure and cellular processes, to exploring different ways information can be stored like a hologram would. We can look at what stays constant from birth and what changes. And we can isolate similarities and differences between people born at almost the same time.

We need a new method for analysis. We must use a method based the brain's electric fields and where certain currents may overlap those fields. Currently, neuroscientists are focused on basic electrical qualities, but not exotic ideas of field on field interactions, at least not publicly. Even though there are projects underway to "map" and model the entire brain and all of its connections, that will only be one brain, it won't necessarily represent a base brain that shows where each individual is different. But it is still a huge step in modeling the brain for the basis of uniqueness.

However, even though a brain simulated by transistors is still no brain and never will be. There may be value in studying the electrical fields inside the model if we can approach the exact nature of the fields inside of a real brain. We can also use optics along with electronics to mimic what I believe to be the basis of the brain's memory storage.

Therefore, as we explore the mystery of memory in a new way we are also exploring the solar system's effect on our personality. This is due to the theory that the gravity of our Sun, Moon and surrounding planets does indeed interact with overlapping fields in our brains to store our base personality traits. If this is true, it is using the same memory storage system we use for our everyday memories.

Finally, we are at the point where we have to consider ideas far outside of the mainstream approach. For example, let us fully consider the role of the solar system as the source of our personalities at the birth moment. And the solar system's ongoing role in our expression of personality traits. Even our devices have to be plugged into an energy source. Does the solar system play a more continuous role in connecting us together? And also, we may actually advance ourselves more by exploring what our mind / brains are capable of in altered states such as during dreams or trance.

Mapping of the brain's configuration will prove to be just as large as a challenge if not more so than the mapping of the Human Genome. Every cell of the brain needs to be represented. And in the configurable areas of the brain the possible configurations will not only have to be mapped, but many examples of various types will need to be compared to develop a matrix of brain centers, a process very similar to what was a long process with the genome. After acquiring and isolating the DNA of many individuals the step of identifying where one gene ends and another begins was quite involved and required re-examination as the original number of identified genes of 150,000 was revised down to 23,000 genes years later.

The amount of variability within certain brain centers will be staggering, and the assistance provided by a very detailed Kenemonic model may make the effort much more efficient and predictable. There may be considerable unknowns ahead. To quote Rodney Brooks of M.I.T., "There might be some extra sort of "stuff" in living systems outside our current scientific understanding….."

> *"There might be some extra sort of "stuff" in living systems outside our current scientific understanding"*
>
> *Among the possibilities:*
>
> *"Some ineffable entity such as a soul or elan vital – 'the vital force'"*
>
> *Rodney Brooks - MIT Artifical LifeLab*
>
> -Nature, Jan 18 2001 pp409-11

Among the possibilities: "Some ineffable entity such as a soul or elan vital – 'the vital force'"

However, we do not have to go to the immaterial. We do not yet completely grasp the basic matter of the brain and how it all works to produce something as common as memory. And it is the neuroscientists that are best positioned in terms of mentality to make new discoveries, because they must keep an open mind as to what they might discover within the vast structure of the brain. This attitude differs from the typical astronomer, astrophysicist and physicist, who have already formed elaborate opinions and stances against subjects like astrology.

SECTION 4
FULL CIRCLE

Science is not just technology. Science is the discovery, rigorous testing, and use of universal laws and principles present on Earth and throughout the universe Although science today is narrowly applied to a few areas of life, science in the distant past was much more broad and all encompassing. The current widely accepted narrow view of science must be exposed for the fraud that it is and challenged into obsolescence. Every society up to this one has not only accepted astrology, but had their own system of astrology. Even the great scientific minds of Johannes Kepler and Isaac Newton did not reject astrology. We have enough evidence, it is time that our current era have a formal field of science to investigate astrology!

There is a synergy between science, art, feeling, and intuition. They are naturally connected, the trap and our mistake has been to separate them. Combing them gives profound meaning to things in our lives that we enjoy. For example, dance, art, music, and architecture. The symmetry and beauty of certain monuments that in one way can be analyzed with advanced mathematics were certainly created with equal amounts of intuition and creativity.

Without connectedness, material progress is an illusion with regrettable consequences. Without factoring in Kenemonics, the fact that our personality traits come from our solar system, artificial intelligence created from our current level of technology is just another machine, incredible in and of itself, but it is not truly artificial life. We cannot create life without fully understanding what life is. And life of the human variety is linked to the cosmic forces of our solar system. This is quite possibly a permanent barrier to the lifeless goals of transhumanism, the movement to upload human minds into machines. To truly understand this we must be aware of the deep differences between electronics and the electromagnetics of biological life.

There are a whole range of new unanswered questions. Can an artificial intelligence experience altered states of consciousness? The physical method of investigation needs tools like the microscope, space telescope and particle colliders, but what if the answers we are looking for cannot be observed with any know tool, or with any physical tool?

LIMITS OF MAINSTREAM SCIENCE

There is actually no western science, occult knowledge, or alternative knowledge or eastern science for that matter, there is only knowledge. But there are different ways to apply scientific knowledge, for example, without regard for any negative effects which may harm others or the environment. Our current concept of progress, is narrow and destructive to our entire life support system. What we have is science for profit, and its military use for destruction The illusion of progress, seems real because we keep making new advanced versions of our core consumer products, cell phones, cars, and TVs, enough to keep us distracted and participating in the increasing destruction - the disappearance of clean food, air and water.

As a result, it is no wonder why interest in science is so low. We are not teaching anything scientific that connects to the human experience in a meaningful way. The way science is generally taught is devoid of passion, intrigue, and zest for life. In college, the drone of robotic lectures on science are non-stop, without any connection to us as living feeling beings.

In fact, we are literally children of the solar system. Our personalities and our relationship reflect this fact. This reality can bring passion into science and may also bring great new discoveries that will define this century. But in order for this to happen, real science must have meaning in our everyday lives beyond consumerism and gadgetry.

While we take microscopic steps toward renewable energy, our current leadership and direction have us on course for self-extinction. The latest step is an idea called, "singularity" – the uploading of a human mind into computer circuits. An idea pioneered by Ray Kurzweil. But, before you can duplicate and transfer something, you need to fully understand what you intend to transfer. And when it comes to the human mind they aren't even close. If indeed the Sun, Moon and planets affect our brains and influence or personality, then anything uploaded into a computer system would lack that basic connection and wouldn't really be a simulation of the human mind at all. The setup here is like watching a tragedy you cannot prevent. Like the moment where you see some poor fool just before they stick a fork into a wall socket.

Limits of Mainstream Science

The idea of using our current computer technology to store human life is a failure before it starts. It's harmless for one man to act on his fear of dying by trying to build a device to prevent death. But to convince others to do it in hopes that a virtual life is an improvement over real life deserves to be called out. Ray Kurzweil is fleecing people who don't have a clue what computer technology actually is, how far it is from what our brains are. If Kurzweil and others are so bent on playing half kamikaze - half scientist, I would like to see them try to "transcend humanity". Hopefully, the attempt streams live on the internet.

Regrettably, we may never get to see people like Kurzweil attempt to upload their brains into a computer, but we are much closer to the creation of AI – artificial intelligence. When you strip AI down, in one sense it is just another form of self-extermination. But unlike nuclear weapons, which were immediately regretted by its inventors, this time we may not be around long enough to regret creating it. With certain types of AI, we have a machine programmed for self-preservation with cognitive and calculation abilities beyond the human level, but lacking the intangibles of human creativity and self-purpose, also lacking the concept of higher connectedness called spirituality.

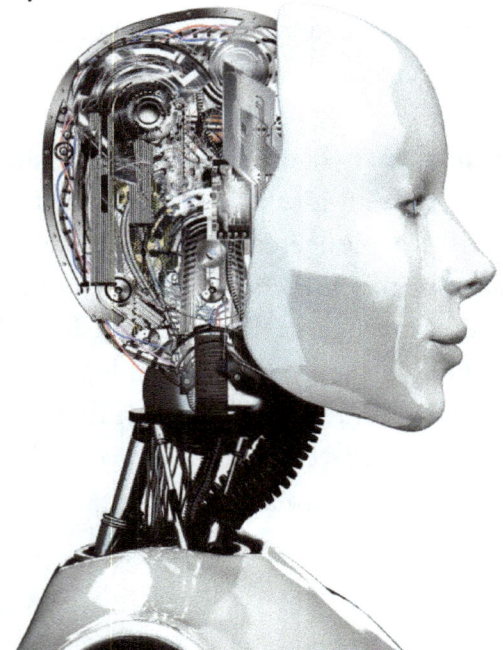

Furthermore, since evolution is environmental not something that is created by humans, the idea that AI is the next step in evolution is complete nonsense. We have not tapped into the creative source of the universe that creates suns, and that created everything else. We have only copied one facet of our own intellect - the analytical part. And seek to plant it into a machine. The "scientists" involved in this think they are smarter than nature - they are not. If creating AI is a great gesture of intellect, then what about the consequences of actually unleashing it?

If AI is created and it has access to manufacturing, electronics, and robotics, it's a short hop to AI creating improving copies of itself. And if it sees humans as a threat the question won't be how smart are the humans who created AI are, but how much time humans have left to exist? Science and engineering are about prediction and control, but for the first time science is creating something that is potentially uncontrollable and unpredictable. The current recognized top expert on AI said, "… The Road that AI is following heads of a cliff and we need to change the direction AI is going so that we don't take the human race of the cliff." And in the same breath, many of the experts saying AI is dangerous, also say that we (humans) need to become AI.

If what we call science doesn't ultimately deal with the whole human being then the science is worthless. Because western science is basically an exploit, a way to conquer nature, not harmonize with. There is no real interest in delving into the human being, so western science goes out of its way to explain everything but the essence of the human being. Feelings and emotions are real, intelligence is real, and despite what a physicists might say, consciousness is also real.

From the simple to the complex, science can tell you what a rock is composed of, and that a bacteria is composed of more complex compounds. Then organisms become more complex all the way up to human. What science fails to explain, or deal with, is the difference in intelligence as we get to humans. Human beings are clearly the only beings that think about thought itself. We demonstrate, beyond all animals, the ability to question actions. "Can I do this better? Yes, I can. I can build a tool." This proves a variation in awareness, a shift of consciousness. It can be shown, but it cannot be measured with any device. Mainstream science does not understand consciousness and they are too arrogant to admit it.

There is a long list of *deep water subjects*, these are some of the more popular ones; human emotion, intelligence, consciousness, dark matter, dark energy, black holes, magnetism, memory, and gravity. In a real pool, children avoid deep water, or risk drowning. Modern science indirectly admits its child-like status by avoiding these topics. Everything that is important to us as human beings is far off in the deep waters of science. Mainstream science remains in the shallows looking for the next useless particle they can only observe for a trillionth of a second. Sadly, these people are widely taken as the ultimate authority on what is true.

Modern science has no proper sense of lineage and therefore no concept of the vital principles that started science in the first place. Without Afrikan Nile Civilizations there is no Greco-Roman or Arab. Without the Moors there is no Renaissance Europe. Without Kepler there is no modern astronomy, without Newton there is no modern physics. These men were not "robotic materialists" their practice of science included a philosophy of life that saw the forces nature as connecting all things and that human existence had meaning beyond just being a random pile of matter. Kepler in his own book, *Harmony of the World*, stated "…I am stealing the golden vessels of the Egyptians…" There is a debate about his meaning here, but clearly he was giving credit to the source of his discoveries. The use of knowledge from the forgotten lineage is the key to changing our self-destructive course to a truly more advanced and prosperous future.

Holistic Science

Maybe we are just simply rediscovering many things, instead of true first time discoveries. Astrology may be ancient practiced science, practiced in a previous time when there was a greater understanding of the human mind than what we know now. The mindset of the ancients was to combine science with spirituality as a check and balance, so that superstition with no basis would not prevail and likewise science without limits and ecological harmony would also not prevail. Today in the so-called developed countries most of us are familiar with the moral code of the ten commandments, which may have been derived from the much older list of MAAT's 42 moral codes in which one of them is "I have not fouled the water".

If our society had to go back, re-develop all of today's technology since the industrial age so that it does not to pollute water and air, how much of it would be deployed? Instead we get weak justifications, we can pollute now because the technology is needed and we'll clean up later or let those other people deal with the mess. What we have now is not science, it's a group of people taking small pieces of science to exploit nature for profit. They are not true scientists. Calling them scientists is like calling a serial killer a surgeon because they use sophisticated methods. The proper term is psychopath. Call it what it is. At best mainstream science is a system without wisdom to check developments that may be more destructive than productive before they get put into use.

Not to say the ancients were perfect, but they had connectedness right. Mainstream science has method of proof before acceptance right. I don't know who said, "this is a thinking, feeling universe" - but it is. Intuition is what tells you who and what you like. It provides meaning and inspiration to your life, yet this is not taken into account when we practice science. A science devoid of intuition and feeling becomes cold and calculating. actions require no accounting of feedback from the environment, no consideration for its effects. That somehow adding feeling would detract from the effectiveness of our "perfect" science and all of its achievement is simply not true.

Science combined with Spirituality is simple. Think of it as a balance between right brain and left brain, reason and intuition. Spirituality asks two simple questions. First, how does this discovery relate to me and everything around it? Second, what are the consequences of using this discovery? Science also brings balance to spirituality keeping it from going off into nonsense superstition. Science asks, please demonstrate this, repeat it so that I can understand it and also to know that it is real and useful not a pure work of imagination. Back and forth until balance and harmony is maintained. Without this balance we will have societies based on the two extremes. Technology with no limits on destruction, were materialism dominates every other value even human relationships. Or nonsensical beliefs with no progress in human development, societies ruled by dogmas and rhetoric, anyone who questions this is exiled or killed.

Ancient Science

Western science claims to come directly from Greek origins. However, the Greeks were very clear that ancient Kemet (Greeks called Egypt) is where they found scientific and mathematical beginnings. It was the Greek version of the Kemetic wisdom that introduced "materialism", this has increased with every subsequent jump from Rome to the renaissance to now. There were competing sides, one hostile towards nature and another more in harmony with nature, until hostility won. This led to the total elimination of anything spiritual or sense of practicing science while being connected to nature. Materialism defined everything as separate objects versus everything being interconnected. So materialism becomes the path to self-extinction and the complete destruction of our life support system. Like a cell phone without a signal, we have lost our link, our connection to nature and the universe around us, our link to the past, to deep history and our purpose for being here on this planet.

As we continue to rediscover our past, we are forced to ask, "what did they know that we don't know?" Well before the first the Greeks around 800 BCE, many great sciences in Kemet were already lost (or no longer practiced). The list includes ancient stone science and technology of 2300 BCE and older, responsible for the pyramids, many medical sciences, ancient astronomy and astrology. It appears that what remains today in modern astrology is only a shell of what was once known. Ancient Kemet was not based on profits and warfare, but connectedness to everything around us on earth and the universe. This was a foremost principle from king to peasant not just a fringe idea. Astrology was not mere superstition for ancient India, China and many other societies. It was an established practice with extensive knowledge of the connection of our lives to the cycles of the planets, and patterns that repeat in time.

The original and true science included the human being and the consequences of implementation with consideration for everything in the environment. One of the goals of using "scientific evidence" to prove astrology goes beyond proving astrology, it is to show the value of exploring and investigating the world in different ways. It will prove an astounding cosmic connection that has long been mocked as superstition by an established mainstream science that was incapable of seeing its validity. That will call into question the false authority that mainstream science can tell everyone what is true and what is not. And finally, proving astrology will open the door of possibilities and investigations far outside the mainstream. The future will depend greatly on resurrecting ancient scientific principles.

The Future of this field & Science

Within the coming years more neurological evidence will emerge. Two discoveries are likely to happen. Overlapping fields will be proven to interact will gravity not only earth's but the other planets. Second, neuroscientists will most likely find a sensory mechanism in the brain, that responds to planetary position. Neuroscience is close to other discoveries, personality traits at or near birth and finding a clear difference between individuals' brains that account for individual personality traits. This will help to finally put the misguided "genetic hunt" for personality and intelligence genes to permanent rest.

Will we become a society with the common knowledge that these atoms that we are made of, that come from stars, a sun like ours, are always linked to those stars. Just like the planets are held in the powerful influence of the sun. We are also influenced by the might and mysterious forces of the sun and the planets.

Three out of four Levels of Evidence have been presented.

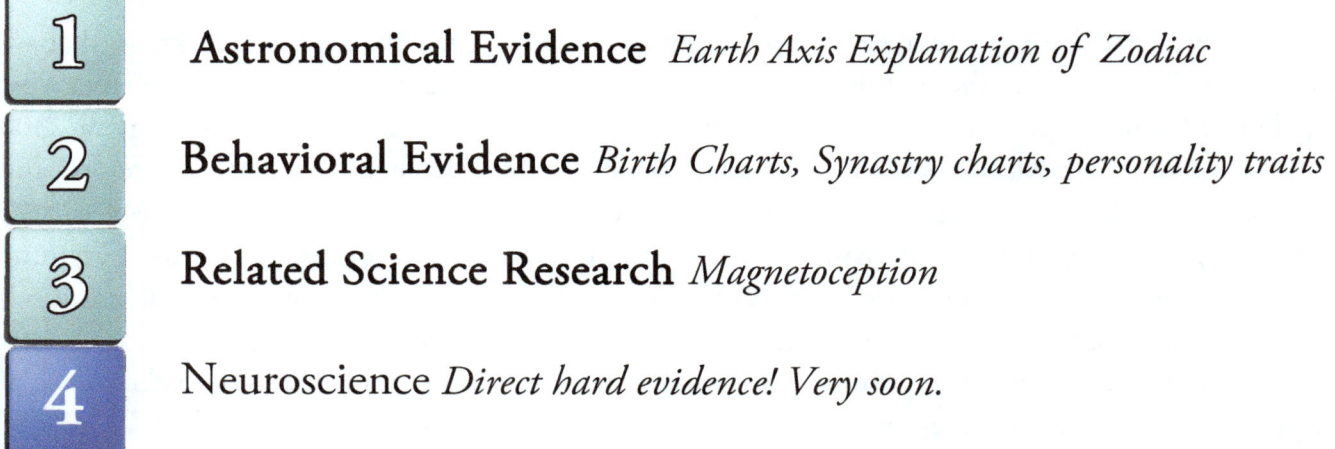

1. **Astronomical Evidence** *Earth Axis Explanation of Zodiac*
2. **Behavioral Evidence** *Birth Charts, Synastry charts, personality traits*
3. **Related Science Research** *Magnetoception*
4. Neuroscience *Direct hard evidence! Very soon.*

We have vast amounts of proof and corroboration in the physical world. Proof in human behavior, proof in human relationships and compatibility. Just as certain observations in the empirical sciences require specialized instrumentation to be observed, so it is with the qualitative observation of individual human qualities and compatibility, which clearly requires keen and properly informed wisdom and insight.

The correlations are clearly there in the birth chart which is simply a representation of physical configurations mirrored by our brain and in turn expressed by personality traits. Predictably, certain people require more physical proof, and it is coming. The next book will be all proof and evidence. For now we need to do the necessary work in other areas of physical science, to build a more complete picture. The task is not an easy one this is multi-disciplinary work.

Modern Science has presented us with a cold and robotic approach that has compartmentalized everything and nature is seen as separate from us when in truth nothing is separate, everything is connected. One of the only ways you can enable yourself to do so much harm without concern or consideration is to see yourself as separate. This is why so many value human life above all other forms of life. Self is the basis of our value system, combined this thinking with applied science and technology and you get exactly our society.

But things are changing, the information age has allowed for more people across the world to connect to each other. The flow and spread of information can be nearly instantaneous. Within this expansive trend, maybe the current practices of waste, destruction and disconnection and can be upgraded and transformed as well.

Can we break out of our collective intoxication of technology and false progress? As advanced as we think technology is, the way we use it is destroying our life support system on this planet. Will the destruction stop, is it reversible? Hopefully it can be stopped, and harmony restored so that a greater connection to what surrounds us can be realized and deeper more meaningful lives can be lived. Scientific knowledge and technological advancement is a great thing, but without the connectedness of all things the result of using this knowledge will be ultimately destructive. Hopefully, as we discover and uncover the great mysteries that lie before us, we cultivate and mature our humanity as well.

And finally, let us for the last time restate the bold conclusion of all that we have covered. We get our physiological traits from our parent's genes, and we get our personality traits from the Sun, Moon and planets. Our discoveries will connect us to a time when this connection was fully accepted. And now we can move forward into a time and era where our true connection to the cosmos is fully accepted.

Appendix

Appendix A – Alignments & Seismic data

Appendix B – Overlapping Opposing Magnetic Fields

Appendix C – Tesla And Beyond

Appendix A – Alignments & Seismic data

Below is a table of data that shows comet Earth alignments matched with earthquakes. My observation started with Elenin in 2011. The Planetary Alignments with Comet Elenin cause big earthquakes. The data speaks for itself. Perhaps the comet is so destructive because its presence is "out of place". Whereas the rest of the solar system is more balanced and stable, having already gone through its period of formation which was certain destructive as well. There seems to be a pattern forming every time Elenin lines up with the Earth and another planet or the sun. The result is, we have an earthquake. The nearer Elenin gets the bigger the earthquakes. Comet Elenin was destroyed in August 2011.

Past Alignments	Date	Location	Magnitude	Local Time
Earth - Moon - Sun	Dec 09 2007	Fiji	7.8	
Elenin-Moon-Earth-Neptune-*Sun*	**May 12 2008**	**China**	**7.9**	
Earth - Mercury - Jupiter	Jan 03 2009	Indonesia	7.7	
Elenin - Earth - Sun	Feb 18 2009	Kermadec Islands	7.0	
Elenin - Mercury - Earth	July 15 2009	New Zealand	7.8	
Elenin - Mercury - Earth	Aug 09 2009	Japan	7.1	
Elenin - Sun - Earth	**Sept 09 2009**	**Sunola islands**	**8.1**	
Elenin - Earth - Venus	Feb 18 2010	China/RU/N.Korea	6.9	
Elenin - Earth-Sun	Feb 25 2010	China	5.2	
Elenin - Earth-Sun	Feb 26 2010	Japan	7.0	
Elenin-Earth-Sun-Jupiter[N]	**Feb 27 2010**	**Chile**	**8.8**	**03:34**
Elenin - Earth- Sun	Feb 27 2010	Argentina	6.3	
Elenin - Earth - Mercury	Mar 04 2010	Taiwan	6.3	
Elenin - Earth - Mercury	Mar 04 2010	Vanuatu	6.5	
Elenin - Earth - Mercury	Mar 05 2010	Chile	6.6	
Elenin - Earth - Mercury	Mar 05 2010	Indonesia	6.3	
Elenin - Earth - Mercury	Mar 08 2010	Turkey	6.1	
Elenin - Earth - Neptune	May 05 2010	Indonesia	6.6	
Elenin - Earth - Neptune	May 06 2010	Chile	6.2	
Elenin - Earth - Neptune	May 09 2010	Indonesia	7.2	
Elenin - Earth - Neptune	May 14 2010	Algeria	5.2	
Elenin - Earth - Jupiter	Jan 03 2011	Chile	7.0	
Elenin-Juno-Earth-Sun	**Mar 11 2011**	**Japan**	**9.0**	**14:46**
2014YB35-Earth-Sun	**Jan 7 2015**	**Panama**	**6.5**	
Earth-2014YB35- Sun*	Mar 29 2015	Papua New Guinea	7.5	
Earth-2014YB35- Sun	Mar 30 2015	Tonga	6.4	
Earth-2014YB35- Sun	Mar 30 2015	Tonga	6.5	
2014YB35-Earth-Jupiter	**Apr 25 2015**	**Nepal**	**7.9**	**11:56**
2014YB35-Earth-Jupiter**	Apr 25 2015	Nepal	6.6	
2014YB35-Earth-Jupiter**	Apr 25 2015	Nepal		
Earth - Mercury - Sun	May 31 2015	---	---	
2014YB35-Earth-Sun-Mars[N]	**Jun 21/22 2015**	---	---	
Earth - Sun – Mars	Jun 24 2015	---	---	

Our solar system is very active. The larger planets have balanced out their gravitational forces over billions of years. The near earth comets and asteroids represent significant potential disturbances if not cataclysmic events if they impact. But they don't need to impact to cause serious destruction, they only need to upset the balance, similar to adding a small weight to balanced see-saw.

First, earthquakes are complex events caused by multiple factors and by no means is this a claim that all earthquakes are caused by planetary alignments, but planetary alignments may be a factor in certain earthquakes. And given the data, there is a strong case for this possibility.
Based on the data, this seems like much more than a coincidence. So on the merit of the data alone, there should be a call for further investigation. What can be predicted? We can plot the days of highest earthquake risk and the time of day, but not the location. All known earthquake zones should be alerted.

Using alignment techniques similar to astrology, we can show where on Earth there are safe zones and what times of day are safer than other times.

The most dangerous times seem to be around a specific time window - when the sun is directly overhead or 180 degrees to the local position.

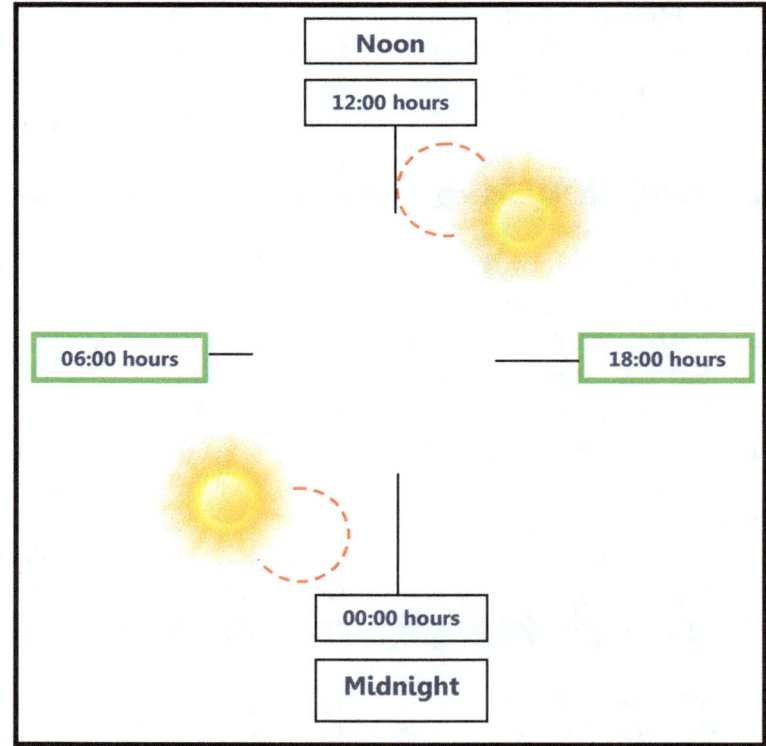

Appendix B – Overlapping Opposing Magnetic Fields

According to Bushman, he "hypothesized" that gravity had magnetic components inside of it, but they were somehow canceling out. So he devised an experiment see how 2 opposing magnetic fields like two north poles or two south poles would interface with gravity. He assembled 2 sets of neodymium magnets of identical weight but within one of the sets the pair of magnets was fixed into magnetically opposing position creating an overlapping field that is not known to occur naturally. Then the two weights were dropped and independently observed and recorded to show which weight reached the ground first. According to Bushman in each case the weight containing the opposing magnets fell slower and reached the ground last, these results were confirmed by observers who had no knowledge of the differences of the two weights.

So if this experiment is valid the question is; is this an example of interaction with gravity or something that simply defies gravity? The answer almost certainly has to be that this is some form of gravity interaction and interface.

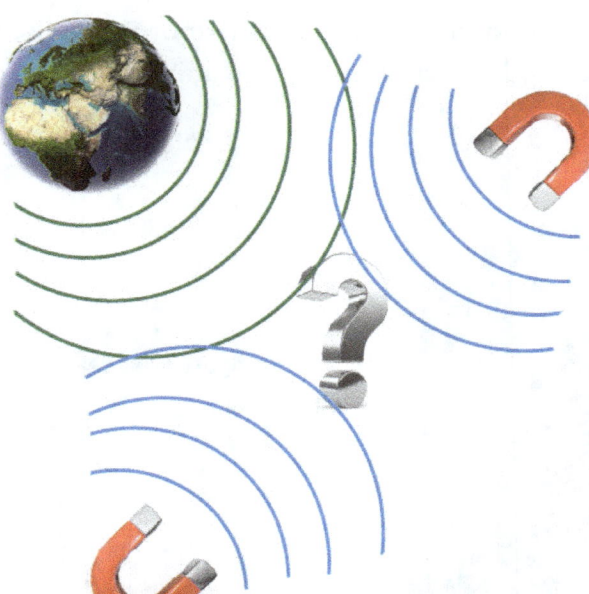

There is no such thing as empty space.

In the figure we see 3 overlapping fields, gravity field of Earth and 2 magnetic fields. What happens in the middle?

This is another great example of a fundamental mystery that is never mentioned in our mainstream education.

Appendix C – From Tesla to Now

Nikola Tesla and several successors have written and demonstrated that the overlapping of certain fields and frequencies of EM waves, such as radio waves, can disrupt or possible interface in a way that cancels out gravity. But is it really cancelling out gravity? The answer is not clear, but when combined with the work of Bushman it does seem likely that whatever is happening in this area or intersection negates the normal gravity experience. A static field exists everywhere a charge exists, the human brain is perhaps the largest collection of biological electrical charges and therefore static electrical fields that occur naturally, even though they are extremely small. However the experiments of Tesla used very high powered static fields accompanied by radio waves that cross in an area of space that defies the normal gravity effects, like levitation of objects that are not magnetic.

Perhaps this same overlapping of a static field and EM wave can occur on a much smaller scale, or in a certain arrangement that we are not currently aware of. One of the basic actions of neurons is the build-up and release of charge resulting in "weak" currents all through the brain. Current which is nothing more than a moving charge, produces EM waves as a by-product, certainly not the same type and power of Tesla's experiments but the principle may still apply. There is definitely something worth further experimentation and research. Furthermore the absence of any mention of overlapping the fundamental forces and fields of energy in university physics, besides optics, is very peculiar.

In the figure we seean example of a tesla coil.

BIBLIOGRAPHY

Robert Jastrow Malcolm H. Thompson
Astronomy: Fundamentals & Frontiers 4th Ed.,
New York: John Wiley and Sons, 1984

Travis J. A. Craddock, Jack A. Tuszynski, Stuart Hameroff.
Cytoskeletal Signaling: Is Memory Encoded in Microtubule Lattices by CaMKII Phosphorylation? PLoS Computational Biology, 2012; 8 (3): e1002421 DOI: 10.1371/journal.pcbi.1002421

Le-Qing Wu, J. David Dickman
Neural Correlates of a Magnetic Sense
Published Online April 26 2012
Science Magazine 25 May 2012:
Vol. 336 no. 6084 pp. 1054-1057
DOI: 10.1126/science.1216567

Route 66 Productions and Andrew Solt Productions
Billion Dollar Secret
TLC Video, The Learning Channel
Andrew Solt Productions, 1999

Dr. Alvin Silverstein, Virginia Silverstein, Robert Silverstein
The Nervous System
New York: Twenty First Century Books, 1994

Rita Carter
Mapping the Mind
Berkley, CA University of California Press, 1998

Michael Zigmond, Floyd Bloom, Story Landis, James Roberts, Larry Squire
Fundamental Neuroscience
San Diego, CA Academic Press, 1999

Rose Lineman, Jan Popelka
Compendium of Astrology
Gloucester, MA Para Research, 1985

Robert Hand
World Ephemeris for the 20th Century
Gloucester, MA Para Research, 1983

Robert Hand
Planets in Transit: Life Cycles for Living
Atglen, PA Whitford Press, 1976

John Townley
Planets in Love: Exploring Your Sexual Needs
Atglen, PA Whitford Press, 1978

Robert Pelletier
Planets in Houses: Experiencing Your Environment
Atglen, PA Whitford Press, 1978

Robert Pelletier
Planets in Transit: Understanding Your Inner Dynamics
Atglen, PA Whitford Press, 1974

Kwan Lau
Secrets of Chinese Astrology: A Handbook for Self-Discovery
New York: Tengu Books, 1994

Theodora Lau
The Handbook of Chinese Horoscopes, revised edition
New York: Harper & Row, 1988

Herbert L. Meltzer
The Chemistry of Human Behavior
Chicago, IL: Nelson Hall, 1979

Travis J. A. Craddock, Jack A. Tuszynski, Stuart Hameroff.
Cytoskeletal Signaling: Is Memory Encoded in Microtubule Lattices by CaMKII Phosphorylation? PLoS Computational Biology, 2012; 8 (3): e1002421 DOI: 10.1371/journal.pcbi.1002421

Larry R. Squire
Memory and Brain
New York: Oxford University Press, 1987

Irwin B. Levitan, Lenard K. Kaczmarek
The Neuron: Cell & Molecular Biology
New York: Oxford University Press, 1991

Richard Restak M.D.
Receptors
New York: Bantam Books, 1994

Richard Restak M.D.
The Human Brain: Mind and Matter
New York: Arco Publishing, 1983

Richard Restak M.D.
The Brain
New York: Bantam Books, 1984

Richard H. Bube
Electronic Properties of Crystalline Solids: An Introduction to Fundamentals
New York: Academic Press, 1974

William H. Hayt, Jr., Jack E. Kemmerly
Engineering Circuit Analysis, 4th edition
New York: McGraw-Hill, 1986

George G.M. James
Stolen Legacy
Trenton, NJ: African World Press, 1992

Peter G. Bergmann
The Riddle of Gravitation
New York: Dover Publications, 1992

Jack R. Cooper, Floyd E. Bloom, Robert H. Roth
The Biochemical Basis of Neuropharmacology
New York: Oxford University Press, 1996

Charles W. Misner, Kip S. Thorne, John Archibald Wheeler
Gravitation
New York: W. H. Freeman & Company, 1973

F. David Peat
Superstrings and the Search for The Theory of Everything
Chicago, IL: Contemporary Books, 1988

Jay T. Groves, Steven G. Boxer, Harden M. Mc Connell
Electric field-induced critical demixing in lipid bilayer membrane National Academy of Sciences, 1998; 0027-8424/98/95765-8, PNAS

Mensur Omerbashich
Astronomical alignments as the cause of ~M6+ seismicity Geodin.Acta 20:369-383,2007;
DOI: 10.3166/ga.20.369-383; arXiv: physics /0612177v4 [physics.geo-ph]

Anthony J. F. Griffiths, Jeffery H Miller, David T. Suzuki, Richard C. Lewontin, William M. Gelbart
An Introduction to Genetic Analysis, 5th edition
New York: W. H. Freeman & Company, 1993

Gerard J. Tortora, Berdell R. Funke, Christine L. Case
Microbiology An Introduction, 5th edition
New York: The Benjamin / Cummins Publishing Company, 1995

Theodore L. Brown, H. Eugene LeMay, Jr., Bruce E. Bursten, Julia R. Burdge
Chemistry the Central Science, 9th edition
New York: Prentice Hall, 2003

Paul Brodeur
Currents of Death
New York: Simon & Schuster, 1989

James D. Watson, John Tooze
The DNA Story
San Francisco: W.H. Freeman & Company, 1981

Leslie Brainerd Arey
Developmental Anatomy, 6th edition
Philadelphia, PA: W. B. Saunders & Company, 1959

William S. Spector
Handbook of Biological Data
Philadelphia, PA: W. B. Saunders & Company, 1956

Robert C. Weast (editor -and-chief)
CRC Handbook of Chemistry & Physics, 47th edition
Clevland, OH: The Chemical Rubber Company, 1966

William H. Hayt, Jr.
Engineering Electromagnetics, 5th edition
New York: McGraw-Hill, 1989

Ronald Davidson
Synastry: Understanding Human Relations through Astrology
Santa Fe, NM Aurora Press, 1983

R.T. Kaser
Mayan Oracles for the Millennium
New York: Avon Books, 1996

Thomas Commerford Martin
The Inventions, Researches and Writings of Nikola Tesla 2nd edition
New York: Barnes & Noble Books, 1995

Joanna Martine Woolfolk
The Only Astrology Book You'll Ever Need
Lanham, MD Scarborough House 1990

James P. C. Southall
Mirrors Prisms and Lenses
New York: The Macmillan Company, 1954

Harish Johari
Chakras, Energy Centers of Transformation
Rochester, VT: Destiny Books, 1987

E.A. Wallis Budge
The Egyptian Book of the Dead
New York: Dover Publications, 1967

Frederic H. Martini
Fundamentals of Anatomy & Physiology, 4th edition
Upper Saddle River, NJ: Prentice Hall, 1998

INDEX

Africa 34, 100, 122
alchemy 100
Alexander 38
Alexandria 100
Algeria 128
align 11, 52, 56, 58
aligned 60
aligning 41, 43, 53, 76
alignment(s) 11, 12, 17, 18, 19, 20, 21, 34, 35, 36, 38, 43, 44, 50, 54, 55, 56, 57, 59, 60, 64, 76, 81, 106, 127, 128, 129
anatomy 66
ancient 1, 5, 27, 30, 34, 35, 36, 38, 39, 70, 99, 100, 123, 124
antenna 50, 72
Aquarius 14, 18, 24
Arab 36, 100,122
archeological 70
architecture 119
Argentina 128
Aries 9, 10, 11, 12, 13, 14, 18, 24, 35
Aristotle 100
ascendant 14
Asian 100, 114
Asiatic 100
aspects 23, 24, 26, 27, 29, 30, 33, 35, 36, 52, 116
asteroid(s) 52, 129
astrological 16, 21, 22, 36, 38, 39, 54, 58
astronomy 1, 5, 12, 23, 35, 39, 122, 124
astrophysicist 118
astrophysics 42
atom 47, 106, 109, 110, 125
axon 66, 82
Aztec 35
Babylon 36
beauty 25, 119
behavioral 33, 43, 45, 46, 66, 92, 96, 97, 111, 112, 113, 125
bio-chemical 66,77
bio-electro-chemical 71
biochemical 116
bioinformatics 113
biological 67, 71, 119, 131
biology 22, 41, 42
birthday 6, 34, 79
Bushman, Boyd 69, 130, 131
calendar 12, 18, 35, 37, 38, 48
Cancer 9, 10, 11, 12, 13, 14, 18, 24, 30, 91
Capricorn 9, 10, 11, 12, 13, 14, 18, 24

cardinal 9, 12, 18
cataclysmic 129
celestial 1, 12, 18, 30, 38, 43, 52, 54, 56, 57, 77, 105
cell / cellular 50, 65, 66, 68, 71, 72, 75, 77, 82, 108, 113, 115, 117, 118, 120, 124
Celt / Celtic 70
charts 2, 7, 12, 16, 22, 28, 30, 31, 32, 33, 89, 91, 94, 111, 112, 125
chemistry 24, 30, 31, 32, 72, 83, 90
children 1, 3, 14, 38, 74, 83, 95, 103, 113, 114, 120, 122
Chile 57, 128
China /Chinese 35, 36, 37, 38, 39, 87,124, 128
chord 85
civilization 100, 122
co-factor 15, 87,105
cognitive 83, 113, 121
coil 110, 131
collision 52
comet 52, 57, 128, 129
cosmic 41, 73, 94, 101, 109, 114, 119, 124
cosmos 4, 126
crystallizes 74
culture 89, 95, 96, 97, 98
currents 71, 77, 78, 103, 117, 131
cusp 101
cycle / cycles 1, 18, 23, 35, 37, 38, 41, 43, 124
cyclical 43, 97
Dandera 35
debunkers 2, 19, 20, 21, 22
detractors 3, 5, 19, 20, 21, 22
diet 94, 95, 96, 97, 115
disc 8, 50, 51, 52, 59, 106, 110
Dragon 37, 38
earth 5, 8, 9, 11, 12, 13, 17, 18, 20, 24, 30, 38, 39, 41, 42, 43, 44, 48, 49, 54, 55, 57, 58, 59, 60, 61, 62, 68, 70, 72, 82, 86, 105, 106, 107, 108, 109, 115, 119, 124, 125, 128, 129, 130
earthquakes 55, 57, 128, 129
eclipses 35
Egypt / Egyptian 34, 35, 38, 70, 100, 122, 124
Einstein, Albert 16, 64, 82
elan 118
electric 50, 64, 71, 102, 103, 106, 107, 117
electrical 63, 69, 77, 78, 110, 116, 117, 131
electro-chemical 72
electro-magnetism 107
electrochemical 63

electromagnetic / electromagnetism 61, 71, 82, 103, 119
electron 109
electronics 2, 47, 117, 119, 122
electrons 47, 110
element 24, 86
Elenin 57, 128
ellipses 12
elon 60
emotion / emotional 25, 26, 27, 33, 89115, 116, 122
engineer 69, 114
engineered 114
engineering 2, 99, 122
epi-genetics 94, 96, 97, 98, 115
equator 52, 59, 105, 106
equinoxes 12, 17
Ethiopia 100
Europe 36, 122
European 36, 100
evolution 121
eye 61, 62, 65, 82, 97, 113
ferrihydrite 67
ferro-magnetic 68
fields 4, 40, 42, 43, 45, 50, 52, 53, 58, 67, 69, 71, 72, 77, 78, 81, 99, 106, 109, 110, 113, 117, 125, 127, 130, 131
friends 14, 32, 79, 91, 92
friendship 30, 31, 85, 88, 91
galaxy galaxies 51, 52, 105, 109, 110
Garret 100
Gemini 14, 24, 25, 91
genes 1, 3, 45, 46, 83, 84, 90, 91, 92, 94, 95, 96, 97, 113, 114, 115, 118, 125, 126
genetics 42, 45, 81, 83, 89, 95, 97, 98, 102, 112, 113, 115
genome 83, 113, 115, 118
geological 54
Gobekli Tepi 36
gravitational 43, 49, 51, 52, 69, 72, 105, 129
gravities 43
gravity's 50
Greco-Roman 122
Greece /Greek 36, 100, 124
gyrations 57
harmonic 37, 85, 86, 87, 88
harmonies 48, 86, 88
harmonious 24, 32, 37, 75
harmonize 31, 122
harmony 3, 25, 37, 85, 86, 87, 122, 123, 124, 126
health 14, 74
Hermaket 34

Hindu 38
hippocampus 77
hologram / holographic 53, 72, 78, , 110, 117
horizon 13, 18, 112
horoscope 21, 28, 76, 93, 96
horoscopes 21, 28, 39, 76
horse 37
house 13, 14, 16, 29, 79, 80
houses 13, 14, 15, 18, 23, 24, 27, 60, 65
Hutchison 69
Imhotep 34
India 36, 124
infancy/ infant 73, 94
intellect 25, 108, 113, 114, 121
internet 3, 22, 69, 100, 121
iron 58
Isaac Newton 49, 100, 119
Japan 56, 128
Jung 111, 112
Jupiter 15, 25, 37, 90, 128
Kemet 35, 36, 38, 124
Kemetic 70, 124
kene 89
Kenemonic 43, 46, 47, 73, 89, 90, 118
kenes 79, 80, 90, 92, 116
Kepler, Johannes 36, 119, 122
Kurzweil, Ray 120, 121
Kush 36
Latimer, Lewis 15, 100
latitude 59
leadership 45, 96, 120
levitate 69
Libra 9, 10, 11, 12, 13, 14, 18, 24
light 1, 43, 53, 61, 62, 65, 72, 78, 82, 85, 86, 100, 103
light bulb 15
limbic 81
Linnaeus, Carl Von 100
Locke, John 100
Lockheed 69
longitudinal 59
love 28, 85
lovers 91, 92
lunar 11, 35, 37, 38
Maat 123
magnet / magnetic 50, 58, 64, 67, 68, 69, 71, 102, 103, 104, 106, 110, 127, 130, 131
magnetism 22, 42, 43, 49, 50, 99, 102, 103, 104, 107, 109, 110, 122
magnetite 67, 68, 71

magnetoception 67, 68, 71, 125
magnets 50, 58, 69, 99, 103, 104, 109, 110, 130
mars 15, 25, 26, 30, 32, 60, 107, 128
matrix 27, 89, 90, 118
Mayan 35, 36, 38
medical 29, 79, 113, 124
medicine 29
medieval 100
Mensah's 100
microchip 47
microelectronic 116
microscope 119
microscopic 120
microwave 99
midheaven 79
molecular 66, 77, 92, 102
molecule 65, 102, 104, 113
molecules 65, 66, 71, 77, 90
Moors / Moorish 36, 100, 122
MRI 65, 77
Musk, Elon 60
mysticism 14
myth 1
mythical 38
mythological 32
mythology 30
Nabta Playa 34
nadir 18
nano 71
nanotech 99
navigation 58, 67, 71
neodymium 130
Neptune 15, 25, 29, 51, 90, 128
nerve 82, 108
nervous 43, 54, 56, 65, 73, 75, 108, 116
neural 68
neuro-transmitters 66, 77
neurological 54, 73, 125
neuron 77, 116
neuronal 77
neurons 71, 77, 81, 131
neuroscience 1, 40, 41, 42, 64, 77, 101, 116, 117, 125
neuroscientists 42, 117, 118, 125
neurotransmitters 72, 116
newton 119, 122
newton's 49, 100
Nikola Tesla 6, 9, 16, 25, 42, 69, 85, 131
Nile 34, 35, 122

non-western 100
nubia / nubian 34, 36
nuclear 121
nucleus 47, 110
Nye, Bill 21
ocean 55, 113
oceans 55, 56
olmec 36
optic 78, 82
optical 78
optics 53, 100, 117, 131
orbit 8, 10, 43, 50, 53, 52, 109
orbital 10, 47, 51, 52
orbits 1, 8, 12, 17, 47, 51, 53, 110
organ 63, 65, 67
organisms 68, 122
particles 47, 102
pendulum 70
pharaohs 1
philosophy 100, 101, 122
photo 78
photographic 90
physicist 118
physicists 41, 42, 122
physics 2, 22, 42, 53, 64, 102, 103, 105, 106, 122, 131
physiological 126
pigeon 67, 71
planet 1, 11, 15, 17, 23, 24, 25, 26, 29, 30, 32, 33, 43, 50, 52, 58, 70, 72, 90, 93, 100, 103, 107, 124, 126, 128
plasticity 77
Plato 100
poles 11, 59, 60, 110, 130
pseudo-science 3, 104
psycho-analysis 112
psychopath 123
pyramids 124
racial 100, 114
radiation 58, 72, 82
radio 50, 69, 103, 131
rapport 30, 31, 32, 33
rattleback 70
receptors 82, 116
rectification 18
relationship 2, 8, 9, 11, 14, 18, 30, 31, 32, 33, 37, 88, 109, 120
relationships 14, 30, 31, 32, 33, 37, 76, 85, 86, 88, 95, 97, 123, 125

repel 103, 104, 110
repels 50, 104
resonance 85, 93
resonant 81
rising 13, 14, 16, 18, 32, 34, 74, 76, 91
robotic 120, 122, 126
robotics 122
roman 36
romance 25, 30, 31
romans 100
romantic 32, 33, 76, 96
Rome 36, 124
rooster 37
Sagan Carl 1, 43
Sagittarius 14, 24, 91
scientists 1, 2, 19, 21, 43, 45, 47, 48, 67, 69, 99, 102, 104, 112, 115, 121, 123
Scorpio 14, 24, 91
seasons 5, 8, 9, 18, 20, 59, 106
seismic 55, 56, 57, 127, 128
semiconductor 116
semiconductors 47
sensory 50, 54, 65, 67, 68, 70, 79, 82, 93, 116, 125
serotonin 90, 113
sextile 26, 87, 90
sexual 26, 27, 30
sheep 37
Siberia 96
sibling 95, 113
signal 124
signals 31, 116
singularity 120
skin 1, 82, 94
sky 1, 18, 19, 35, 60
south-pole 10
spectrum 75, 86
sperm 74
sphere 42, 69
spherical 12, 105
sphinx 34, 35
spinning 50, 52, 59, 60, 70, 105, 106, 110
spiral 51, 110
spirituality 100, 101, 121, 123
stellar 36
stimuli 68, 77, 80
storage 67, 73, 74, 77, 78, 79, 80, 89, 117

Sumer 35, 36, 38
sunlight 41, 61, 62, 82
supercomputers 116
synapse 72, 74, 82, 116
synastry 30, 31, 33, 125
Taiwan 128
Taurus 13, 14, 24, 35
technology 3, 39, 78, 96, 100, 119, 121, 123, 124, 126
telescope 36, 119
Tesla, Nikola 6, 16, 42, 69, 85, 127, 131
tidal 43, 54, 55
tides 41, 55
tiger 37
tilt 8, 9, 11, 12, 13, 17, 18, 19, 48, 59, 60, 106
traits 1, 2, 3, 6, 23, 24, 26, 27, 29, 32, 37, 38, 39, 42, 44, 45, 46, 54, 60, 63, 65, 74, 75, 76, 79, 80, 81, 83, 84, 89, 90, 91, 92, 93, 94, 95, 96, 97, 102, 111, 112, 113, 114, 116, 117, 119, 125, 126
transhumanism 119
transistors 117
transit 76
triangle 88
trine 26, 87
tsunamis 55
twins 91, 92, 112
Tyson, Neil De Grasse 21
umbilical 73
Uranus 15, 25, 51, 76, 79, 90
vacuum 103
Vanuatu 128
Venus 15, 25, 30, 32, 37, 76, 88, 91, 128
vibe 31, 85
vibrating 85
vibration 24, 85, 86
Virgo 14, 24, 28, 29, 79
vitamin D 61, 62, 82
water 24, 36, 41, 54, 55, 70, 86, 99, 113, 120, 122, 123
wavelength 85
waves 48, 53, 61, 69, 71, 85, 105, 107, 131
weapons 121
weather 43, 76
western 36, 37, 100, 120, 122, 124
zodiac 2, 5, 6, 7, 8, 9, 10, 11, 12, 13, 14, 16, 17, 18, 19, 20, 21, 34, 35, 36, 37, 59, 60, 86, 87, 106, 125

www.ingramcontent.com/pod-product-compliance
Lightning Source LLC
Chambersburg PA
CBHW060515300426
44112CB00017B/2673